Management and Industrial Engineering

Series Editor

J. Paulo Davim, Department of Mechanical Engineering, University of Aveiro, Aveiro, Portugal

This series fosters information exchange and discussion on management and industrial engineering and related aspects, namely global management, organizational development and change, strategic management, lean production, performance management, production management, quality engineering, maintenance management, productivity improvement, materials management, human resource management, workforce behavior, innovation and change, technological and organizational flexibility, self-directed work teams, knowledge management, organizational learning, learning organizations, entrepreneurship, sustainable management, etc. The series provides discussion and the exchange of information on principles, strategies, models, techniques, methodologies and applications of management and industrial engineering in the field of the different types of organizational activities. It aims to communicate the latest developments and thinking in what concerns the latest research activity relating to new organizational challenges and changes world-wide. Contributions to this book series are welcome on all subjects related with management and industrial engineering. To submit a proposal or request further information, please contact Professor J. Paulo Davim, Book Series Editor, pdavim@ua.pt

More information about this series at https://link.springer.com/bookseries/11690

Carolina Machado
Editor

Challenges and Trends in Organizational Management and Industry

Springer

Editor
Carolina Machado
School of Economics and Management
University of Minho
Braga, Portugal

ISSN 2365-0532 ISSN 2365-0540 (electronic)
Management and Industrial Engineering
ISBN 978-3-030-98047-4 ISBN 978-3-030-98048-1 (eBook)
https://doi.org/10.1007/978-3-030-98048-1

This Springer imprint is published by the registered company Springer Nature Switzerland AG
The registered company address is: Gewerbestrasse 11, 6330 Cham, Switzerland

Preface

Competitiveness and change are very common concepts in today's organizations. In order to become competitive, organizations must assume a proactive stance that, rather than responding to market demands, contributes to an effective anticipation of the needs of different stakeholders who directly or indirectly interact with organizations. With increasingly educated and informed costumers, organizations make significant efforts to pursue higher levels of flexibility, quality, creativity and innovation. Considered as one of the most distinctive factors at the organizational level, the organization' human resources assume a fundamental role here.

Endowed with a wide range of knowledge, skills and competences, the organization' human resources provide the fundamental support for promoting the desired creativity and innovation. Similarly, along with technological innovation, it is crucial to implement continuous organizational innovation. Indeed, the potential of technological innovation can only be achieved if the organization prepares itself by having the necessary material, financial and, above all, human resources able to "transform" the organization into a more efficient and effective exercise of its activity. A more decentralized organizational structure and a more dynamic, participative, organic-flexible and more conscious management are critical to cope with the high levels of competitiveness that organizations are increasingly facing.

From the foregoing, it is clear that there are many the challenges that organizations are facing not only today but increasingly in the approaching times. In fact, today and more and more in the future, the key to survival and the organizational success is no longer based on rational and quantitative approaches, but on the involvement of something more "irrational", difficult to measure, as are, for example, the human resources, the quality of customer service and, most importantly, the development of flexibility to cope with change.

What challenges and trends are felt in the market, and what are the implications for their operation, are therefore a key concern facing these organizations, which the book proposed here seeks to address.

Drawing on the input of a wide range of researchers and scholars from management and organizational engineering, this book, entitled *Challenges and Trends in*

Organizational Management and Industry, aims to highlight some of the key challenges that today's organizations are facing, as well as the trends that are gradually being felt within organizations, and their impacts on their functioning.

Following these concerns, this book looks to cover the field of *Organizational Management and Industry* in six chapters. This book focuses on "Employees' Behavioural Response to Organisational Change: Effect of Psychological Contract and Change Readiness"; speaks about "Practising Servant Leadership: Pastoral and Social Ministry in the Modern Age"; highlights "Homosexual Collaborators' Perception About Discrimination by the Organization and/or Peers: A Case Study"; contains information about "Corporate Social Responsibility: From the Origin to the COVID-19 Pandemic"; deals with "Interviewer Characteristics and Nonresponse Survey Outcomes: A Portuguese Case Study"; finally covers "Balanced Scorecard to Improve Human Capital Management: Case Study in a Portuguese Company".

Challenges and Trends in Organizational Management and Industry can be used by a variety of potential stakeholders, including academics / researchers, managers, engineers, practitioners and other professionals in the different areas of business and management. A very special target audience consists of students from different undergraduate / graduate levels (undergraduate, masters and doctoral/PhDs), from the areas of management and industrial engineering, for which this book constitutes a fundamental support capable of providing a focused and current view about the key challenges, trends, implications, strategies and ways of overcoming, that dynamic and competitive organizations are facing today and increasingly in the near future.

The Editor acknowledge her gratitude to Springer for this opportunity and for their professional support. Finally, we would like to thank to all chapter authors for their interest and availability to work on this project.

Braga, Portugal — Carolina Machado

Contents

Editor and Contributors

About the Editor

Carolina Machado received her Ph.D. degree in Management Sciences (Organizational and Policies Management area / Human Resources Management) from the University of Minho in 1999, Master degree in Management (Strategic Human Resource Management) from Technical University of Lisbon in 1994, and Degree in Business Administration from University of Minho in 1989. Teaching in the Human Resources Management subjects since 1989 at University of Minho, she is since 2004 Associated Professor, with experience and research interest areas in the field of Human Resource Management, International Human Resource Management, Human Resource Management in SMEs, Training and Development, Emotional Intelligence, Management Change, Knowledge Management and Management/HRM in the Digital Age/Business Analytics. She is Head of the Human Resources Management Work Group at the School of Economics and Management at University of Minho, Coordinator of Advanced Training Courses at the Interdisciplinary Centre of Social Sciences, Member of the Interdisciplinary Centre of Social Sciences (CICS.NOVA.UMinho), University of Minho, as well as Chief Editor of the International Journal of Applied Management Sciences and Engineering (IJAMSE), Guest Editor of journals, books Editor and book Series Editor, as well as reviewer in different international prestigious journals. In addition, she has also published both as editor/co-editor and as author/co-author several books, book chapters and articles in journals and conferences.

Contributors

Simon Okechukwu Ayogu Institute of Social Sciences, University of Minho, Braga, Portugal

Cláudia Cunha Communication and Society Research Centre, Institute of Social Sciences, University of Minho, Braga, Portugal

Aditi Jamil Department of Employment Relations and Human Resource Management, Griffith University, Gold Coast, Australia

Regina Leite School of Economics and Management, Interdisciplinary Centre of Social Sciences (CICS.NOVA.UMinho), University of Minho, Braga, Portugal

Carolina Machado Department of Management, School of Economics and Management, Interdisciplinary Centre of Social Sciences (CICS.NOVA.UMinho), University of Minho, Braga, Portugal

Carla Magalhães The Transdisciplinary Research Center of Innovation & Entrepreneurship Ecosystems (TRIE), Universidade Lusófona Do Porto, Porto, Portugal

Alice Delerue Matos Department of Sociology, Institute of Social Sciences, University of Minho, Braga, Portugal

Pedro Novo Melo School of Technical Short Cycles, Polytechnic Institute of Cávado and Ave, Braga, Portugal;
Interdisciplinary Centre of Social Sciences—CICS.NOVA.UMinho, Braga, Portugal

Wayne O'Donohue Department of Employment Relations and Human Resource Management, Griffith University, Gold Coast, Australia

Jéssica Perestrelo School of Management, Polytechnic Institute of Cávado and Ave, Barcelos, Portugal

Katrina Radford Department of Employment Relations and Human Resource Management, Griffith University, Gold Coast, Australia

João Leite Ribeiro School of Economics and Management, Interdisciplinary Centre of Social Sciences (CICS.NOVA.UMinho), University of Minho, Braga, Portugal

Liliana Isabel Mendes Vilarinho School of Economics and Management, University of Minho, Braga, Portugal

Gina Voss Communication and Society Research Centre, Institute of Social Sciences, University of Minho, Braga, Portugal

Matt Xerri Department of Employment Relations and Human Resource Management, Griffith University, Gold Coast, Australia

Employees' Behavioural Response to Organisational Change: Effect of Psychological Contract and Change Readiness

Aditi Jamil, Wayne O'Donohue, Matt Xerri, and Katrina Radford

Abstract The purpose of this study is to understand the effect of employees' psychological contract and change readiness on employees' behavioural responses towards organisational change. This research uses a theoretical framework which links employees' response to organisational change based on employees' psychological contract type and level of individual change readiness. Four hypotheses are proposed and tested using a quantitative approach. Data were collected through an online survey that was completed by 117 aged care employees working in residential and community care in Australia. The study results reveal that employees with predominantly relational psychological contracts and high levels of individual change readiness are most likely to respond through Voice when organisational change occurs. On the other hand, employees with predominantly transactional psychological contracts and low levels of individual change readiness are most likely to respond through Neglect when organisational change occurs. As organisations are constantly facing change, it is important for successful change implementation to understand how employees are likely to respond to intended, unintended or unpleasant change. By understanding the link between perceived non-fulfilment of psychological contract and individual's change readiness level, managers can design and implement change actions which will enhance the likelihood of a positive and constructive response to change necessary for organisational and employee performance.

Keywords Employees' Behavioural Response · Organisational Change · Psychological Contract · Change Readiness

1 Introduction

Organisational change is an ongoing area of interest for management scholars and practitioners alike, as organisations are constantly undergoing change processes. From an organisation's development perspective, organisational change is the

A. Jamil (✉) · W. O'Donohue · M. Xerri · K. Radford
Department of Employment Relations and Human Resource Management, Griffith University, Gold Coast, Australia
e-mail: aditi.jamil@griffithuni.edu.au

C. Machado (ed.), *Challenges and Trends in Organizational Management and Industry*, Management and Industrial Engineering,
https://doi.org/10.1007/978-3-030-98048-1_1

purposeful effort to alter work processes with the objective of enhancing organisational performance, as well as contributing to the development of individual employees (Porras & Robertson, 1992). However, within the strategic management literature, organisational change is the process of corporate strategy implementation (Dunphy, 2000). However way change is defined, it is now commonly accepted that all organisations are subject to multiple and continuous change. Therefore, it is essential to explore organisational change to guide organisations and employees through this process.

Previous research demonstrates that an employee's psychological contract is an antecedent to the successful acceptance and implementation of change initiatives (Bouckennooghe, 2010; Jones et al., 2005). Prior empirical research also demonstrates that organisational change can directly influence employees' perceptions and expectations about their post-change relationship with the organisation. In other words, organisational change influences, and is influenced by, an employee's psychological contract (Schalk et al., 1998). According to Schalk et al. (1998), the psychological contract shapes employee attitudes, and attitudes are related to employee behaviour.

Armenakis (1993) indicates that change readiness is an antecedent to the behaviours of either resistance to, or support for change initiatives. Vakola (2014) argues that readiness for organisational change is one of the most important factors relating to employee support for organisational change initiatives, and Stevens (2013) highlights the crucial role of individual change readiness in determining the outcome of any change initiative. However, to the best knowledge of the authors, there is no extant evidence for the interaction between psychological contract type and individual change readiness as an influence on employee behavioural responses to change. Given this lack of evidence, this study examines the joint effects of these variables in the context of organisational change in the aged care community.

2 Theoretical Background

2.1 Psychological Contract

Rousseau (1990) defines a *psychological contract* as, "the individual's beliefs about mutual obligations, in the context of the relationship between employer and employee" (p. 391). In this sense, the psychological contract comprises an individual employee's beliefs, shaped by the employing organisation, regarding terms of an exchange agreement between the individual and the organisation where beliefs refer to the employee's interpretation of explicit and implicit promises made to the employee by the organisation (Rousseau, 1995). Psychological contracts focus on employee perceptions of *organisational obligations* (that are perceived to be obligatory for the organisation to fulfil) and *employee obligations* (that are perceived to be obligatory for the employee to fulfil), based on the promises made by the employer and employee, respectively (Van der Smissen et al., 2013).

Building upon the earlier work of MacNeil's (1985) typology of contracts, Rousseau (1995) describes the difference between two basic types of psychological contracts: (i) *transactional psychological contracts* are normally time-bound (i.e. operate over a finite period of time) and involve exchange between the employer and employee that is mainly economic in nature, whereas (ii) *relational psychological contracts* are not necessarily time-bound and involve a relationship between the employer and employee that is based mainly on socio-emotional exchange. It is important to note here that both forms of currency—economic and socio-emotional—are exchanged under both types of contract, i.e. what varies is the predominance of each currency according to type. Rousseau (1995) elaborates on these differences by stating that transactional contracts have well-specified performance terms, whereas relational contracts have abstruse performance requirements. Regarding the relationship between predominant psychological contract and employee characteristics, Rousseau (1995) states that transactional contract employees have low membership commitment and weak identification with the organisation, have higher freedom to enter new contracts and have simpler Exit responses. On the other hand, relational contract employees have high membership commitment, high affective commitment and high identification with the organisation (Rousseau, 1995).

In a further elaboration of the two basic psychological contract types, Thompson and Bunderson (2003) outline an 'ideology-infused' psychological contract type, i.e. a variation of either of the two basic types based on the premise that "employee perceptions about the organisation's obligations are not grounded solely in personal entitlements but also in the promotion of a cause they highly value" (Thomson & Bunderson, 2003, p. 571). An ideology-infused contract involves the exchange, not only of economic currency and socio-emotional currency, but also of another type of currency—ideological currency—between employer and employee. In psychological contracts involving the exchange of ideological currency, employees believe they are engaged in meaningful work that has a societal purpose. In such cases, the organisation's perceived obligations include the ongoing pledge to support a credible social cause, while the employees perceived obligations include supporting this social cause through both organisational and societal citizenship (Thomson & Bunderson, 2003).

Generally, psychological contracts are shaped by the employment context; Shore et al. (2004) note the key influence of the employment setting or what can be considered as the terms and conditions (both explicit and implicit) of the relationship between the employer and employee. Shore et al. (2004) further state that the importance of employment setting, which can be complex and multi-dimensional, is that it provides both opportunities and limitations that affect the perceived obligations between employee and employer; for example, "the individual in a traditional job might be embedded in nested contexts involving the job, workgroup, department, division, organisation, occupation, labor market, family and national culture. Within the organisation, social, physical, and structural aspects of context might also influence the relationship" (Shore et al., 2004, p. 325).

Since the employment context is an important factor in shaping the relationship between employee and employer, it can be expected that when the organisation undergoes change the employment setting and exchange relationships that comprise the psychological contract will also change. On this point, Freese (2007) refers to organisational change as efforts to reorganise the employment setting in order to develop and improve the performance of both the organisation and employees, and presents evidence that employee perception of non-fulfilment by the organisation of its obligations under the psychological contract frequently accompanies such efforts. The development of perceptions of non-fulfilment can influence employee behaviour negatively (Rousseau, 1995), as well as prompt a re-calibration of employee psychological contracts (e.g. Chaudry et al., 2011).

2.1.1 Types of Behavioural Response (Exit, Voice, Loyalty and Neglect)

The Exit-Voice-Loyalty-Neglect framework, conceptualised by Hirschman (1970) and further developed by Farrell (1983) and Rusbult et al. (1988), has a long history of use in psychological contract research to understand employee response to organisational change (e.g. Chaudry et al., 2009; O'Donohue et al., 2015). These four types of responses can be characterised using two dimensions: active vs. passive and constructive vs. destructive. The active/passive distinction is based on how direct and effortful the behavioural response is, whereas the constructive/destructive distinction is based on how the behavioural response positively or negatively affects the employee-organisation relationship (O'Donohue et al., 2015).

According to Rousseau (1995, p. 135), "Exit is voluntary termination of the relationship"; it is a form of active, destructive response. Exit includes leaving a job, thinking about leaving a job, transferring to other duties/positions and looking for new employment opportunities (Rusbult et al., 1988), and is most likely to happen in employment relationships which are self-regulating and of relatively brief duration (Rousseau, 1995). Voice refers to employees making active attempts to improve a situation (Farrell, 1983), including raising issues with colleagues and supervisors, identifying solutions to problems and trying to implement these solutions to the mutual benefit of the organisation and the employee. This is an active, constructive response to organisational change and is most likely to happen when a long-term relationship and trust exist between the employee and the organisation (Rousseau, 1995). Loyalty refers to the employee response of "passively but optimistically waiting for conditions to improve" (Rusbult et al., 1988, p. 601). Chaudhry et al. (2009) elaborate that Loyalty involves supporting the organisation, demonstrating patience and continuing organisational citizenship behaviours by the employee. Rousseau (1995) terms Loyalty as a passive, constructive response through which employees wait optimistically for conditions to improve. Neglect is a passive but destructive response because it represents "lax and disregardful behavior" (Farrell, 1983). Types of Neglect behaviours include absenteeism, lateness, poor work performance, etc., which are reflective of the individual's disregard for the relationship with the organisation (Farrell, 1983).

2.2 Individual Change Readiness

In addition to psychological contract type, the other variable of interest here is *individual change readiness*, in other words the individual employee's attitude towards organisational change. Holt et al. (2007) define individual change readiness as "… a comprehensive attitude that is influenced simultaneously by the content (i.e. what is being changed), the process (i.e. how the change is being implemented), the context (i.e. circumstances under which the change is occurring), and the individuals (i.e. characteristics of those being asked to change) involved" (p. 235). Others such as Choi (2011) emphasise the importance of the individual's beliefs regarding the need for change and their ability to make the change: "readiness is also concerned with the extent to which an individual believes that a change at the individual level is needed and whether he or she has the capacity for it" (p. 481).

Change readiness, in the form of a willingness and ability to accept change, is essential to the successful implementation of organisational change as organisational goals are achieved through the collective efforts of individual employees (Jones et al., 2005). As noted by George and Jones (2001), "organisations only change and act through their members, and even the most collective activities that take place in organisations are the result of some amalgamation of the activities of individual organisational members" (p. 420). Employees can have varying levels of individual change readiness ranging from high to low. Research demonstrates that individual differences in change readiness relate to how employees respond differently when experiencing organisational change (Oreg et al., 2011; Vakola et al., 2014). Vakola (2014) indicates there is a positive relationship between an individual's core self-evaluation, which is a personality trait that is based on how individuals assess themselves and their relationship with the environment (Judge et al., 1998), and an individual's change readiness level. Core self-evaluation is comprised of self-esteem, self-efficacy, locus of control and emotional stability (Judge et al., 1997). Individuals who have high change readiness have positive attitudes and are proactively ready to support change (Vakola, 2014). Individuals who have high change readiness are also more confident of the success of change implementation (Vakola, 2014), and they may welcome change, as it can be a way to gain benefits for themselves (Judge et al., 1999). However, employees who have low change readiness are not likely to behave proactively and are likely to reject novel work practices as a threat that offer neither benefits nor improved status (Vakola, 2014). Hence, employees with low change readiness may have negative attitudes towards change.

3 Research Framework and Hypotheses

Schalk et al. (1998) conclude that employees' perception of psychological contract fulfilment by the organisation is a key variable for an organisational change implementation process to succeed. Chaudhry et al. (2009) suggest that in times of change, where change is perceived as non-fulfilment of psychological contract by the organisation, there is less likelihood of employees displaying Loyalty behaviour and a greater likelihood of Exit, Voice and Neglect behaviour. On the other hand, where change is not perceived as non-fulfilment, Loyalty behaviour is a greater likelihood than Exit, Voice and Neglect behaviour (Chaudhry et al., 2009).

At the same time, according to Robinson et al. (1994), the type of psychological contract (transactional or relational) perceived by the employee can moderate a negative employee behavioural response to organisational change. Taking into account type of psychological contract, O'Donohue et al. (2015) suggest that employees with a predominantly relational psychological contract will more likely respond to non-fulfilment by the organisation through constructive behaviours, either passive (Loyalty) or active (Voice); by definition, such employees will have a long-term employment relationship based on socio-emotional obligations and trust with high affective commitment to the organisation (Rousseau, 1995). When there is an event that is perceived as non-fulfilment of psychological contract, such employees are willing to be constructive in their response through co-operative behaviours aimed at restoring their relationship with the organisation (O'Donohue et al., 2015). In contrast, employees with a predominantly transactional psychological contract will more likely respond through destructive behaviours, either passively (Neglect) or actively (Exit). Since such employees have a relationship with the organisation based primarily on financial exchange and self-interest (Rousseau, 1995), behavioural responses to perceived non-fulfilment by the organisation are more likely to be destructive and negatively impact on such employees' relationship with the organisation (O'Donohue et al., 2015).

Drawing on this research, Fig. 1 demonstrates the conceptual model utilised in this study:

Based on this conceptual framework, the following four hypotheses are proposed:

Hypothesis 1: When an organisational change is perceived as non-fulfilment of PC, employees with a predominantly relational psychological contract and low change readiness are more likely to respond through Loyalty.

Hypothesis 2: When an organisational change is perceived as a non-fulfilment of PC, employees with a predominantly relational psychological contract and high change readiness are more likely to respond through Voice.

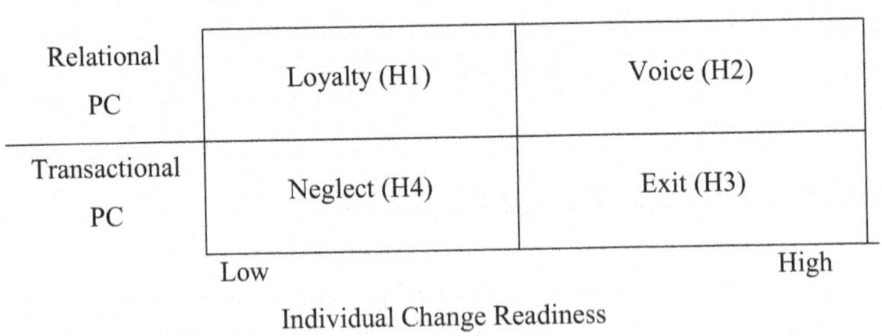

Fig. 1 Relating psychological contract types and level of change readiness to EVLN responses

Hypothesis 3: When an organisational change is perceived as a non-fulfilment of PC, employees with a predominantly transactional psychological contract and high change readiness are more likely to respond through Exit.

Hypothesis 4: When an organisational change is perceived as a non-fulfilment of PC, employees with a predominantly transactional psychological contract and low change readiness are more likely to respond through Neglect.

4 Methodology

4.1 Sampling

The sample has been drawn from aged care employees working in Australia including both residential care and community care settings. The sample included 117 participants from one large organisation based in South Australia. Participants were predominantly female (87.7%) and over the age of 45 (77.2%). Most of the participants were from residential aged care facilities (62.3%) and were holding the position of a personal care worker (26.3%). In terms of employment, there was a mix of full-time (47.7%) and part-time staff (42.5%), as well as staff who have ongoing (61%), contract (29.2%) and casual (9.7%) employment. This reflected the various types of employments existing in the aged care industry.

4.2 Measures

Psychological contract types: The relational and transactional psychological contracts were measured based on the shortened scale originally developed by Millward and Hopkins (1998) which included seven items to measure relational type psychological contract ($\alpha = 0.70$) and ten items to measure transactional type psychological contract ($\alpha = 0.70$) (see Raja et al., 2004, for discussion of the psychometric properties of the scale). Participants were asked to respond to a five-point Likert type scale ranging from 1 = 'strongly disagree' to 5 = 'strongly agree' in response to each statement (item). The scores for relational psychological contract type and transactional psychological contract type were computed by averaging the items that are related to that particular type of psychological contract. Sample items associated with transactional PC type included, "I prefer to work a strictly defined set of hours" and "I expect to benefit from any extra effort I put in". Sample items associated with relational PC type included, "I feel part of a team in this organisation" and "I feel this organisation reciprocates the effort put in by its employees".

Individual's level of change readiness: An individual's level of change readiness was measured based on the scale developed by Vakola (2014). The items were found to load clearly on one factor (Vakola, 2014) and the reliability of the scale ($\alpha =$

0.70) was also acceptable (Vakola, 2014). This scale includes six items. Similar to Vakola (2014), a seven-point Likert type scale (1 = 'strongly disagree' and 7 = 'strongly agree') was used to measure participants' agreement with each item. The average responses to the six items reflected the overall score for an individual's level of change readiness.

Behavioural responses: The behavioural intentions of Exit, Voice, Loyalty and Neglect were measured based on a scale developed by Naus et al. (2007). Naus et al. (2007) tested the convergent and discriminant validity of these scales. Naus et al. (2007) also demonstrated that the composite reliabilities of each of these four scales were higher than the recommended cut-off of 0.70 (Nunnally & Bernstein, 1994). The Exit subscale had a composite reliability of 0.90, while the composite reliabilities for the Voice, Loyalty and Neglect scales were 0.86, 0.87 and 0.90, respectively. Participants were asked to provide their responses on a five-point scale ranging from 1 = 'strongly disagree' to 5 = 'strongly agree'. There were 20 items in the Naus et al. (2007) scale measuring potential behavioural responses: five items each for measuring Exit, Voice, Loyalty and Neglect. Scores for each of the scales were computed by averaging the responses to the relevant items for each scale.

5 Data Analysis and Results

5.1 Data Analysis

For this study, multivariate analysis of variance (MANOVA) was identified as the most appropriate statistical technique, as MANOVA helps to determine whether there are any differences between independent groups on more than one dependent variable (Conlon, 2016). A one-way analysis of variance (ANOVA) was performed on each behavioural response type with the four groups of PC types and level of change readiness, following which MANOVA was performed to compare the four groups of PC type and level of individual change readiness on the four behavioural responses to change (EVLN).

In order to assess the validity of the scales measuring psychological contract, change readiness and behavioural responses to change (Exit, Voice, Loyalty and Neglect), all 44 items related to these variables were included in an exploratory factor analysis utilising varimax rotation and a force rotation of seven factors. As the study was expecting more than five factors, a force rotation was performed (Conlon, 2016).

The final scales included: seven items for transactional psychological contract; seven items for relational psychological contract; five items for change readiness; and twenty items for behavioural response. The factor loadings of the items on their respective scales for the analysis were all higher than 0.40 and none of these items had a cross-loading > 0.40 on any two factors. Hence, the scales in the study all demonstrated convergent validity. Moreover, the factor correlation matrix in Table

1 shows that the correlations between factors were less than 0.7, suggesting that discriminant validity was present in the analysis.

To assess the reliability of each scale, the Cronbach's alpha for each of the scales was calculated (see Table 1). The calculations suggested that the reliabilities of all of the scales were higher than 0.70, indicating that all the scales demonstrated acceptable reliability (Hair et al., 2006). For this study, the reliability score for transactional and relational psychological contract scales was $\alpha = 0.77$ and $\alpha = 0.88$, respectively, which are similar to the reliability scores of the subscales used by O'Donohue et al. (2015). The reliability of the individual change readiness scale ($\alpha = 0.86$) was also close to the scale developed by Vakola (2014). The Exit subscale had a composite reliability of 0.91, while the reliability for the Voice, Loyalty and Neglect scales were 0.85, 0.87 and 0.90, respectively, and exactly the same as the scale developed by Naus et al. (2007).

In order to determine respondents' predominant psychological contract type (PC type), respondents' scores on transactional psychological contract and relational psychological contract were standardised and the dimension with the highest average score was then used. Each respondent's predominant psychological contract type was then coded accordingly as a nominal value where 1 represented predominantly transactional psychological contract (PreTran PC) and 2 represented predominantly relational psychological contract (PreRel PC). Hence, if the transactional psychological subscale score was higher of the two standardised scores, the respondent's psychological contract type was classified as PreTran PC. On the contrary, if the relational subscale score was higher of the two standardised scores, the respondent's psychological contract type was classified as PreRel PC.

Respondents' change readiness scores were calculated by using the average responses to each of the five items to determine the overall score. Classification of respondents as having either high or low change readiness was done using a mean split. Each respondent's change readiness score was coded as a nominal variable, with scores at or below the mean of all respondents' change readiness scores ($M = 5.21$) signifying low change readiness and scores greater than the mean signifying high change readiness.

In order to create the required combinations of psychological contract type and level of change readiness, respondents' predominant psychological contract type and change readiness level were joined to form a four-group categorical variable: (a) PreTran PC with low CR, (b) PreTran PC with high CR, (c) PreRel PC with low CR and (d) PreRel PC with high CR. In terms of the behavioural response to change scale items, the average responses for the five items used for each response type were used to calculate the overall score for each behavioural response type. Table 2 provides the number of respondents in each group.

Table 1 Means, Standard Deviation, Correlations and Reliabilities

	Mean	SD	Cronbach's Alpha	Transactional PC	Relational PC	Change readiness	Exit	Voice	Loyalty	Neglect
Transactional PC	2.41	0.73	0.77	1						
Relational PC	3.09	0.84	0.88	−0.39**	1					
Change readiness	5.18	1.00	0.86	−0.40**	0.53**	1				
Exit	3.03	0.94	0.91	0.16	−0.45**	−0.26**	1			
Voice	3.89	0.67	0.85	−0.42**	0.40**	0.49**	−0.28**	1		
Loyalty	2.66	0.79	0.87	0.03	0.24**	0.17	−0.24**	0.07	1	
Neglect	1.85	0.88	0.90	0.33**	−0.25**	−0.34**	0.19*	−0.21*	0.12	1

*Correlation is significant at the 0.05 level (2-tailed) ** Correlation is significant at the 0.01 level (2-tailed)

Table 2 Groups Based on Psychological Contract Type and Level of Change readiness

	N	Percentage
PreTran PC with low CR	38	32.48
PreTran PC with high CR	11	9.40
PreRel PC with low CR	26	22.22
PreRel PC with high CR	42	35.90

5.2 *Results*

Analysis of the survey responses demonstrated that 42% of the respondents had a PreTran PC whereas 68% of respondents had a PreRel PC (see Table 1). Survey responses demonstrated that 55% of the respondents had a low level of change readiness, whereas almost 45% had a high level of change readiness. The survey analysis also revealed that 78% of the respondents with PreTran PC have low change readiness, whereas 62% of respondents with PreRel PC had high change readiness.

Analysis showed that PreRel PC was positively related to change readiness, Voice and Loyalty. PreRel PC was negatively related to Exit and Neglect. PreTran PC was negatively related to change readiness and Voice. PreTran PC was positively related to Neglect. Among the behavioural response variables, Exit was negatively related to Voice and Loyalty, as well as being positively related to Neglect. Voice and Neglect were also negatively correlated. The MANOVA results in Table 3 revealed a statistically significant pattern of responses between the four groups of PC type and level of individual change readiness at $p < 0.05$; Wilks's lambda = 0.70 (F = 0.46); Pillai's trace = 0.31 (F = 3.24); Hotelling's trace = 0.41 (F = 3.66).

The ANOVA results (see Table 1) showed that the differences between the four groups were significant ($p < 0.05$) on only one of the four behavioural response types,

Table 3 ANOVA and MANOVA Results

Dependent Variables	Group Mean (SD)						Homogeneity Variables	
	PreTranPC with Low CR	PreTranPC with High CR	PreRelPC with Low CR	PreRelPC with High CR	F	P	Levene Statistics	Sigf
Exit	3.35(0.87)	3.04(1.20)	2.85(0.87)	2.85(0.92)	2.37	0.07	1.19	0.32
Voice	3.52(0.59)	3.93(0.95)	3.75(0.54)	4.29(0.51)	11.45	0	3.75	0.01
Loyalty	2.53(0.78)	2.80(1.42)	2.89(0.66)	2.59(0.65)	1.31	0.28	4.22	0.01
Neglect	2.13(0.97)	1.85(1.34)	1.85(0.74)	1.58(0.66)	2.69	0.05	2.28	0.08
Multivariate Test								
Wilks's lambda = 0.70							3.46	0
Pillai's Trace = 0.31							3.24	0
Hotelling's Trace = 0.41							3.66	0

Voice (Hypothesis 2: $F = 11.45$), and approaching significance on one other response types, Neglect (Hypothesis 4: $F = 2.69, p = 0.08$). However, no significant differences between the groups were observed on the responses, Loyalty (Hypothesis 1) and Exit (Hypothesis 3). Thus, PC type and individual change readiness significantly influenced respondents' behavioural responses towards Voice but did not influence behavioural response towards Exit or Loyalty.

6 Discussion and Contribution

The main research question for this study is *how do employees' psychological contract and change readiness interact to influence employee behavioural responses to organisational change?* Overall, the results of this study suggest that to implement change successfully and to achieve positive change outcomes, organisations need to understand psychological contract types and individual change readiness, and their significance for change implementation. Specifically, the results demonstrate that employees in the aged care industry who have a PreRel PC and a high level of change readiness are more likely to express behavioural responses towards the organisational change through Voice (Hypothesis 2) and employees with a PreTran PC and low level of change readiness are most likely to respond through Neglect (Hypothesis 4), and these results are consistent with the findings of previous research that links PC type to behavioural response (e.g. O'Donohue et al., 2015; Robinson et al., 1994; Rousseau, 1995).

Contrary to the theoretical arguments, this study did not find any significant evidence to support Hypothesis 1 (Loyalty) and Hypothesis 3 (Exit). In recent years, researchers have found new evidence regarding the impact of ideology on the relationship between the employee and the organisation, i.e. employees' psychological contract. Thompson and Bunderson (2003) state that "employees' perceptions about organisation's obligations are not grounded in personal entitlements, but also in the promotion of a cause they highly value" (p. 571). They have defined this as 'ideology', which can act as a factor in shaping the employee-employer/organisation relationship. Employment in the aged care industry may have a component of idealistic service (i.e. through helping the aged population) embedded in it. This can make the psychological contract a tripartite rather than a dyadic relationship. In such an employment contract, it is possible that employees may form multi-party psychological contracts (employee/employer/other, e.g. client or patient), rather than a simple two party relationship with the organisation (employee/employer). Hence, a perception of obligations to the third party which override obligations to the employer may influence the behavioural response of such employees to a perception of non-fulfilment of the psychological contract (Radford & Stevenson, 2018). This makes it possible, in the case of the participants in this study, i.e. workers in aged care organisations, that a commitment to their clients rather than to the organisation may explain the non-significant result in relation to Loyalty as a response, remembering that in this research context Loyalty was defined as towards the organisation. This explanation may also explain a reluctance to respond through Exit, which is an active-destructive response towards the organisation that may affect their clients more directly and negatively.

Another possible reason for the non-support of Hypothesis 3 (Exit) could be the demographics of the sample. As 77% of respondents are aged above 45, they are reluctant to respond through Exit as this implies termination of work. As Kira and Klehe (2016) reveal, job loss for mature aged workers may threaten their self-identity, and hence, they may be reluctant to exit from their current job when alternative employment options are perceived as limited and/or less desirable. Moreover, a decision to exit from a job can be influenced by personal and work circumstances (Perera et al., 2015). In the case of mature age workers, there may be an awareness that they have limited time to achieve their personal goals, and accordingly be more careful about leaving or retiring from work (Perera et al., 2015). Older workers may sometimes also hold on to their job to fulfil criteria for retirement benefits (Beehr et al., 2000).

In summary, the study findings suggest that the effectiveness of organisational change implementation might be significantly affected by employees' psychological contract type and individual change readiness level. To implement organisational change successfully, organisations thus should focus on finding ways to influence these positively before and through the change process.

7 Limitations and Future Research Direction

The small sample size and limiting of respondents to a single industry have implications for the findings, data analysis and generalisability of the study. For example, given the small overall sample size, the four individual groups constructed for the analysis were also small, thus making it difficult to compare responses within the groups and test the hypotheses conclusively. Moreover, a focus on employees working in the aged care industry may have produced a general level of homogeneity in terms of motivational factors and personality types, potentially impacting the generalisability of the study findings. In other words, drawing participants from a number of industries, where the possibility of tripartite psychological contracts was less likely, and where employees focus on employer obligations with regard to employee benefits and rewards, might have produced different results. These limitations offer opportunities for future research that builds on and extends the findings of this study, particularly in relation to tripartite psychological contract types and the exchange of ideological currency. Additional research also needs to be conducted on identifying the factors that affect the level of individual readiness to organisational change.

In conclusion, this study's findings provoke further thought on the hitherto unexamined relationship between psychological contract and change readiness, and which may interact to impact on how employees respond behaviourally towards organisational change. The study provides evidence suggesting that there is value in examining responses to organisational change as a function of the employment relationship and a range of other individual differences, such as psychological capital, locus of control and personality. Also, by understanding more about the links between individuals' change readiness level, perceived non-fulfilment of psychological contract and employee behaviour, managers can design and implement change strategies and actions that enhance employees' change readiness, strengthen the employment relationship and thus make more likely the positive and constructive response to change that is needed for organisational and employee performance.

References

Armenakis, A. A., Harris, S. G., & Mossholder, K. W. (1993). Creating readiness for organisational change. *Human Relations, 46*(6), 681–703.

Beehr, T. A., Glazer, S., Nielson, N. L., & Farmer, S. J. (2000). Work and nonwork predictors of employees' retirement ages. *Journal of Vocational Behavior, 57*(2), 206–225.

Bouckennooghe, D. (2010). Positioning change recipients' attitudes toward change in the organisational change literature. *The Journal of Applied Behavioural Science, 46*(4), 500–531.

Chaudhry, A., Wayne, S. J., & Schalk, R. (2009). A Sensemaking Model of employee evaluation of psychological contract fulfilment. *The Journal of Applied Behavioural Science, 45*(4), 498–520.

Chaudhry, A., Coyle-Shapiro, J. A. M., & Wayne, S. J. (2011). A longitudinal study of the impact of organisational change on transactional, relational, and balanced psychological contracts. *Journal of Leadership and Organisational Studies, 18*(2), 247–259.

Choi, M. (2011). Employees' attitudes toward organisational change: A literature review. *Human Resource Management, 50*(4), 479–500.

Conlon, E. G. (2016). *Using statistics for research: A workbook for issues in quantitative research.* Griffith University.

Dunphy, D. (2000). *Embracing paradox: Top-down versus participative management of organisational change, a commentary on Conger and Bennis* (pp. 123–135). Harvard Business School Press.

Farrell, D. (1983). Exit, Voice, Loyalty, and Neglect as responses to job dissatisfaction: A multidimensional scaling study. *Academy of Management Journal, 26*(4), 596–607.

Freese, C. (2007). *Organisational change and the dynamics of psychological contracts: A longitudinal study.* Tilburg University.

George, J. M., & Jones, G. R. (2001). Towards a process model of individual change in organisations. *Human Relations, 54*(4), 419–444.

Hair, J. F., Jr., Black, W. C., Babin, B. J., Anderson, R. E., & Tatham, R. L. (2006). *Multivariate data analysis* (6th ed.). Pearson Prentice Hall.

Hirschman, A. O. (1970). *Exit, Voice, and Loyalty: Responses to decline in firms, organisations, and states, 25.* Harvard University Press.

Holt, D. T., Armenakis, A. A., Feild, H. S., & Harris, S. G. (2007). Readiness for organisational change: The systematic development of a scale. *The Journal of Applied Behavioural Science, 43*(2), 232–255.

Jones, R. A., Jimmieson, N. L., & Griffiths, A. (2005). The impact of organisational culture and reshaping capabilities on change implementation success: The mediating role of readiness for change. *Journal of Management Studies, 42*(2), 361–386.

Judge, T., Locke, E., & Durham, C. (1997). The dispositional causes of job satisfaction: A core evaluations approach. *Research in Organisational Behavior, 19*(3), 151–188.

Judge, T., Locke, E., Durham, C., & Kluger, A. (1998). Dispositional effects on job and life satisfaction: The role of core evaluations. *Journal of Applied Psychology, 83*(1), 17–34.

Judge, T. A., Thoresen, C. J., Pucik, V., & Welbourne, T. M. (1999). Managerial coping with organizational change: A dispositional perspective. *Journal of Applied Psychology, 84*(1), 107.

Kira, M., & Klehe, U. (2016). Self-definition threats and potential for growth among mature-aged job-loss victims. *Human Resource Management Review, 26*(3), 242–259.

MacNeil, I. R. (1985). Relational contract: What we do and do not know. *Wisconsin Law Review*, pp. 483–525.

Millward, L. J., & Hopkins, L. J. (1998). Psychological contracts: Organisational and job commitment. *Journal of Applied Psychology, 28*(16), 1530–1556.

Naus, F., van Iterson, A., & Roe, R. (2007). Organisational cynicism: Extending the exit, voice, loyalty, and neglect model of employees' responses to adverse conditions in the workplace. *Human Relations, 60*(5), 683–718.

Nunnally, J. C., & Bernstein, I. H. (1994). *Psychometric theory* (3rd ed.). McGraw-Hill.

O'Donohue, W., Martin, A., & Torugsa, N. (2015). Understanding individual responses to failure by the organisation to fulfil its obligations: Examining the influence of psychological capital and psychological contract type. *Human Resource Management Journal, 25*(1), 131–147.

Oreg, S., Vakula, M., & Armenakis, A. (2011). Change recipients' reactions to organisational change: A sixty-year review of quantitative studies. *Journal of Applied Behavioural Science, 47*(4), 461–524.

Perera, S., Sardeshmukh, S., & Kulik, A. T. (2015). In or out: Job exits of older workers. *Asia Pacific Journal of Human Resources, 53*(1), 4–21.

Porras, J. I., & Robertson, P. J. (1992). *Organisational development: Theory, practice, and research.* Consulting Psychologists Press.

Radford, K., & Stevenson, A. (2018). *Extending the models of employee turnover to include external drivers such as client and industry factors: A multi-country, qualitative assessment*, Academy of Management Proceedings.

Raja, U., Johns, G., & Ntalianis, F. (2004). The impact of personality on psychological contracts. *Academy of Management Journal, 47*(3), 350–367.

Robinson, S. L., Kraatz, M. S., & Rousseau, D. M. (1994). Changing obligations and the psychological contract: A longitudinal study. *Academy of Management Journal, 37*(1), 137–152.

Rousseau, D. (1990). New hire perspectives of their own and their employer's obligations: Study of psychological contracts. *Journal of Organisational Behavior, 11*(5), 389–400.

Rousseau, D. (1995). *Psychological contracts in organisations: Understanding written and unwritten agreements*, Sage Publications.

Rusbult, C. E., Farrell, D., Rogers, G., & Mainous, A. G. (1988). Impact of exchange variables on exit, voice, loyalty, and neglect: An integrative model of responses to declining job satisfaction. *Academy of Management Journal, 31*(3), 599–627.

Schalk, R., Campbell, J. W., & Freese, C. (1998). Change and employee behavior. *Leadership and Organisation Development Journal, 19*(3), 157–163.

Shore, L. M., Tetrick, L. E., Taylor, M. S., Coyle Shapiro, J. A. M., Liden, R. C., McLean Parks, J., & Rousseau, D. (2004). *The employee-organisation relationship: A timely concept in a period of transition.* Research in personnel and human resources management (pp. 291–370), Emerald Group Publishing Limited.

Stevens, G. W. (2013). Toward a process based approach of conceptualizing change readiness. *The Journal of Applied Behavioural Sciences, 20*(10), 1–28.

Thomson, J. A., & Bunderson, J. S. (2003). Violations of principle: Ideological currency in the psychological contract. *Academy of Management Review, 28*(4), 571–586.

Vakola, M. (2014). What's in there for me? Individual change readiness and the perceive impact of organisational change. *Leadership and Organisational Development Journal, 35*(3), 195–209.

Van der Smissen, S., Schalk, R., & Freese, C. (2013). Organisational change and the psychological contract: How change influences the perceived fulfilment of obligations. *Journal of Organisational Change Management, 26*(6), 1071–1090.

Practising Servant Leadership: Pastoral and Social Ministry in the Modern Age

Simon Okechukwu Ayogu, João Leite Ribeiro, and Regina Leite

Abstract Changes in societies, individuals and organizations call for new approaches of leadership. Servant leadership has emerged as a value-based leadership style informed by service. Greenleaf (2002, 2003, 2007) presented it as a better alternative to the traditional approaches of leadership in which the leaders exhibit an overriding and determining posture over and above their followers. The present chapter explores the understanding and practice of servant leadership in the Church and its possible impacts on the ministry of the Church and today's Christian community. The data collection method consisted of 15 semi-structured interviews with Church leaders and lay faithful Christians who work in Europe, America and Africa. The findings suggest that servant leadership can be applied to different contexts, cultures and organizations. According to the study participants, it can help today's Church mission and pastoral leaders improving their ministry effectiveness and also serve as a role model for lay professionals in their roles and organizations.

Keywords Servant leadership · Service · Church · Humanity · Teachings of Christ

1 Introduction

Leadership and management are two important matters in every organization and all sorts of human interaction (Fairholm, 2015). As an institution, the Church is also faced with the challenges of leading and managing people, and expected to meet the changing expectations of societies and individuals (Aubrey, 1998; Mittal & Dorfman, 2012; White & Acheampong, 2017). New times require new leaders and methods to

S. O. Ayogu
Institute of Social Sciences, University of Minho, Braga, Portugal

J. L. Ribeiro (✉) · R. Leite
School of Economics and Management, Interdisciplinary Centre of Social Sciences (CICS.NOVA.UMinho), University of Minho, Braga, Portugal
e-mail: joser@eeg.uminho.pt

R. Leite
e-mail: rleite@eeg.uminho.pt

C. Machado (ed.), *Challenges and Trends in Organizational Management and Industry*, Management and Industrial Engineering,
https://doi.org/10.1007/978-3-030-98048-1_2

manage the challenges of the twenty-first century, with Church being no exception (Christensen-Salem et al., 2021; Kahler, 2000; Sun, 2013; White & Acheampong, 2017). Christians in different parts of the world, directly or indirectly, are noticing some changes in their perception of Christianity, expectations from the clergy, and the way they want to be led (Setley & Gautsch, 2015). There is also an increasing diversity among religious groups and egocentricity because of civilization, modernization, a sweeping technological advancement (Liew, 2017; Lobl, 2018; Logan, 2011) and other external stimuli. Leaders in many religious organizations seem to be paddling with opening the Church's doors to a generation that does not seem to value much or understand serving and being led in the way of Christ (Lobl, 2018). Some community members seem to expect the Church leaders to be perfect or clearheaded (Fulenwider, 2004; Greenleaf, 2007). Extant studies argue that the youth wish their leaders were more patient, more communicating and more entertaining, with the elderly more concerned about fostering spiritual growth among worshippers (Agli et al., 2015; Agyei, 2019; Gesme & Wiseman, 2010; Malone & Dadswell, 2018; Ruiz-Prada et al., 2021; Salas-Vallin et al., 2020). Following these lofty expectations, some pastors have introduced significant, disruptive changes that challenged conventional ways of Church management and operations, in their bids to foster more effective pastoral leadership or maintain relevance in the changing world (Zehndorfer, 2013). Given the above scenario, it is vital that the Church, in pursuit of sustainable styles of leadership capable of reawakening the conscience of humanity about the needs of followers (Baker, 2014), increases the number of leaders who devote themselves totally to the service of God and humanity (Embree, 2016).

Servant leadership and other moral-based leadership theories (such as ethical and authentic leadership) have received growing attention over the last decade from both academics and practitioners, due to the positive consequences on relevant individual and organizational outcomes, namely organizational commitment and citizenship behaviour, job satisfaction and performance (Canavesi & Minelli, 2021; Christensen-Salem et al., 2021). Servant leadership pays particular attention to ethics, virtues and morality and is concerned with the well-being of followers including employees, customers and community (Eva et al., 2019; Reddy & Kamesh, 2016). One of the defining features of servant leadership is that its primary objective is to serve and meet the needs of others (McMinn, 2001; Russell, 2001). It has been presented as a leadership style that relies upon the influence of self-giving without self-glory (Stone et al., 2003), prioritizing the fulfilment of followers' needs above the leader's personal needs. In this regard, servant leadership entails inversion of the pyramid of power portraying the leader as one who serves (Athal et al., 2021; Kimotho, 2019). Although coined in the 1970s by Robert K. Greenleaf, servant leadership remains largely understudied, probably due to the fact that it was firstly conceived as a way of life, and consequently difficult to operationalize and apply (Parris & Peachey, 2013). Research on servant leadership is in a relatively early stage, compared to traditional leadership approaches, but existing research shows promising empirical findings (Christensen-Salem et al., 2021; Dierendonck et al., 2014; Keith, 2014).

Modern age Churches face increasing expectations regarding their leaders and members compared to those of the medieval age, specifically on their ability to effectively render services to the followers. The aim of this chapter is twofold: to examine the concept and practice of servant leadership in the Church and the possible impact on the ministry of the Church and Christian community. Another purpose is to ascertain whether the Church in its leadership has become more dynamic and if/how this dynamism helps changing the lives of the people for the better.

The remainder of this chapter is structured as follows. The next section presents a brief outline of the literature on the concept of servant leadership, Greenleaf's conceptualization of servant leadership, servant leadership in the context of religious organizations and the ten servant-leader characteristics developed by Spears (2010) from Greenleaf's writings. The research methodology and a description of the participants are presented in the subsequent section. Next, the interviewees' opinions and perceptions regarding the main themes are presented and discussed. In the final section, the main conclusions are drawn from the empirical study and some suggestions are offered for future research and practice.

2 Literature Review on Servant Leadership

The concept of leadership is one of the most studied in the field of management and organizational behaviour, mainly due to the supposed impact on individuals' performance and organizational success (Armstrong, 2006; DeRue & Ashford, 2010; Ling et al., 2008; Talya et al., 2019). Over the years, different approaches and theories of leadership have emerged, in an attempt to explain this complex phenomenon and its main consequences for individuals and organizations. These theories of leadership are mainly concerned with who the leader is and what the leader does (Parris & Peachey, 2013). In the 1970s, servant leadership enters the organizational context through the seminal works of Greenleaf (1970, 2002). This new paradigm builds upon a concept that echoes the beliefs and messages of ancient and current religious and historic leaders, and teachings of Jesus Christ (Parris & Peachey, 2013).

The servant leadership theory is centred on the idea that at the top of the hierarchy or system are the employees, and that a manager or leader's top priority is to serve them. It is based also on the idea that authority is more important than power, and that focus on empowering and uplifting others instead of personal gain is critical to everyone's success (Alm & Guttormsen, 2021; Christensen-Salem et al., 2021; Leibensperger, 2015; Parris & Peachey, 2013; Savel & Munro, 2017). This theory of leadership argues that the most effective leaders are servants of their people. Servant leaders get results for their organization through whole-hearted attention to their followers and followers' needs (Burkus, 2010). Unlike many approaches to leadership, which offer suggestions on how top-level leaders can influence and motivate those further down the hierarchy, servant leadership puts its emphasis on collaboration, trust, empathy and ethics. The leader should be a servant first, leading from a desire to better serve others and not to attain more power. The assumption is that if leaders focus on the needs and desires of followers, followers will reciprocate through increased teamwork, deeper engagement and better performance (Leibensperger, 2015; Walumbwa et al., 2008).

Servant leadership is gaining a global acceptance and has roots in both Eastern and Western philosophy (Burkus, 2010). Though it is not yet adopted widely, servant leadership is being applied in various organizations in different parts of the world including well-known organizations such as 7-Eleven, TD Industries, Southwest Airlines, Herman Miller and ServiceMaster (Ingram, 2016). While not widely adopted, it is being accepted in many organizations as an ideal option especially for those committed to organizational growth and development (Salas-Vallina et al., 2020).

Regarding the development of servant leaders, many scholars have taken different positions as to whether servant leaders are born or made. Gardner (1990) argues that leaders are born and not made, but he also found that it is reasonable to believe that servant leadership can be taught because leadership itself can be taught (see also Aline & Ramkumar, 2018; Khan et al., 2016; Swaroop & Prasad, 2013). Rost (1991) suggested that a number of leadership concepts make the assumption that leaders are made, not born, but he pointed out that this could be a too narrow view. In his essays, Greenleaf (1970) noted that leaders must learn how to view the act of leading as an act of serving followers. So, leaderships generally hinge on what you can do and not solely or particularly on what you are.

Servant leadership is one of the more popular theories of leadership, especially among Christian leaders who vigorously cite Jesus as the penultimate example of servant leadership (Burkus, 2010). However, its effectiveness in organizations is still being debated, calling for more empirical research in order to overcome the criticism of being predominantly anecdotal in nature (Northouse, 2019). Many researchers and theorists argue that servant leaders can become so focused on the needs of their followers that the needs of the organization suffer as a result (Berson & Avolio, 2004; Setley & Gautsch, 2015). In any case, servant leadership theory has a place within the spectrum of leadership theory, as it represents the strongest emphasis on followers of any theory, being depicted as a servant-first leadership paradigm (Parris & Peachey, 2013).

2.1 Greenleaf's Perspective of the Servant as Leader

The majority of research on servant leadership has developed from the seminal works of a former AT&T executive, Robert K. Greenleaf. His foundational writings have proven to inspire and sustain a meaningful research interest in leadership as service, namely in the works of Spears (1995, 1998). Greenleaf sustains that a servant leader is first a servant and, afterwards, a leader and only leads because he serves, which should be adopted as the path to follow by the Church in her pastoral and social approaches. This choice to serve first, as opposed to the one who is first and foremost a leader, manifests itself in the care taken by the "servant-first" to make sure that other people's highest priority needs are being served (Greenleaf, 2007).

In accordance with Greenleaf's postulation, a servant leader is someone who naturally serves first and then with time, if necessary, makes a conscious choice and aspires to leadership. He does not serve as a leader because service is expected but leads as a servant because leadership is necessary. Service here becomes mutually beneficial to both the server and the served (Chiniara & Bentein, 2016).

The modern Church took its form in a period when, outside the home, the Church was the dominant influence in people's lives and it shaped and sustained the influence of the home (Greenleaf, 2003; Spears, 2010). During that period, it was taken for granted that the Church had the absolute truth about almost everything and this could be seen in the respect which accrued from this in favour of Church figures. In present times, it is becoming abundantly clear that the Church may no longer enjoy the unanimity of followership it had. The schools and workplaces seem to be taking the centre stage when it comes to what marks the rhythm of life for the contemporary individual. As stated in the introduction of this chapter, the Church, as most of the contemporary institutions and organizations, is facing unprecedented changes that Church leaders and members are expected to address. Servant leadership has emerged as a better approach to leadership (Greenleaf, 1970, 2002), based on a true desire to help and serve others.

2.2 *The Place of Servant Leadership in Religious Organizations*

The subject of leadership dwells at the centre of the life of any organization (Greenleaf, 2003) as the servant leaders emphasize follower development (Hale & Fields, 2007). Servant leadership, though not limited to Church organizations, is often associated with Christian leadership and Christian leaders (Ingram, 2016). The concept has been associated with religious beliefs and rooted in Judeo-Christian theology (Rachmawatia & Lantub, 2014; Sendjaya et al., 2008).

As a religious organization, the Church has had many leaders, from the medieval ages to the contemporary age, and has changed as many times as it was deemed necessary (Rego & Cunha, 2008; Rego et al., 2015). The Church has been a beacon of light to the hopeless and has also made some mistakes in her attempts at self-leadership and as a forerunner in many human endeavours including leadership (Lamm, 2021). This seems to corroborate the thoughts of many early Church fathers that "the Church has come to bless, and all its tenets are for human upliftment" (Samuel & Sugden, 1987, p. 10).

Following the evolution of time, the Church has gradually ceased from being considered as a seemingly all-knowing and omnipotent institution (John Paul II, 1991) to a more dynamic one. Nowadays, it is becoming evidently clear that the Church has gradually become a place where the Christians can meet, worship, pray, obtain spiritual guidance and socialize. Yet in her wisdom, she has held a pivotal role in charting the course of humanity on almost all fronts: arts, sciences, humanities, economics, etc. The changes would have telling consequences in the way she perceives her leadership role in society and is perceived by other institutions and the whole world at large (Taylor et al., 2000).

Servant leadership demonstrates that leadership is at its best when it seeks to promote the interests of the followers (Greenleaf, 2007; Rego & Cunha, 2008). The Church will be seen as a system that recognizes leadership as service, (Webster, 2003), an arrangement best illustrated with the classic organizational pyramid turned upside-down.

2.3 *Characteristics of Servant Leadership: A Brief Overview*

Spears (1998) identified ten attributes of servant leaders from Greenleafs' writings: listening, empathy, healing, consciousness, persuasion, conceptualization, forecasting, stewardship, commitment to people growth and community building (Greenleaf, 2002; see also Russell & Stone, 2002). According to Spears, this set of characteristics is critical to the growth and development of servant leaders.

2.3.1 Listening

Listening is the foundation of communication and the key to building strong relationships (Zisa, 2013). Listening and understanding are great leadership attributes. In his essays, Greenleaf (2007) claims that by listening a leader does not have to be, in every case, the bearer of a possible problem (Greenleaf, 2007; Northouse, 2019; Spears, 2010) but gets information necessary to help in solving problems. Many problems actually fade away significantly when all the parties involved decide to listen and more so, when the leader is the chief listener. In this respect, Greenleaf (2002, p. 8) says the following:

> "Why is there so little listening? What makes this example so exceptional? Part of it, I believe, with those who lead, is that the usual leader in the face of a difficulty tends to react by trying to find someone else on whom to pin the problem, rather than by automatically responding: "I have a problem. What is it? What can I do about my problem?" The sensible person who takes the latter course will probably react by listening, and somebody in the situation is likely to say what the problem is and what should be done about it". Or enough will be heard that there will be an intuitive insight that resolves it (Greenleaf, 2007; Spears, 2010). Ideally, in leadership, we serve by creating a space where individual voices can grow and be heard, and where fear-based leadership is overcome (Northouse, 2019).

2.3.2 Empathy

In line with Greenleaf's thought, a servant leader is expected to be able to recognize, value and strive to understand the world view, strength, aspirations, feelings and emotions of the others (Greenleaf, 2007; Wilson, 2017). Such a leader will care for other people and will deeply experience emotions that match what others are feeling, adapting his or her own communication style to meet the needs and expectations of the other person. As the servant leaders understand others so deeply, their actions would be motivated by a genuine desire to help others (Fields et al., 2015).

St. Paul advocates for this virtue of empathy as he wrote in his letters to the Church in Thessalonica: "Therefore encourage one another and build one another up, just as you are doing" (1 Thessalonians 5:11).[1] The life of Jesus Christ reflects so much true servant leadership. He came as a man; he came seeing things with men's eyes, feeling things with men's feelings, and thinking things with men's minds. Seen in this sense, God knows what life is like, because He was able to take the aspect and habits of human nature except sin (Greenleaf, 2003). This shows Jesus Christ is not remote, detached and disinterested, nor insulated and isolated from our lives. He knows our frame; He knows that we are but dust and can see in us a reflection of what He experienced as a man. He can thus extend mercy to us, completely understanding what we are going through. Therefore, becoming a skilled empathetic listener is of utmost importance for the success of servant leaders (Tran & Spears, 2020), as portrayed by Christ.

2.3.3 Healing

Followers typically desire for a leader who has a sincere interest in fostering their emotional and spiritual well-being (Wilson, 2017). The potential for healing those who suffer is presented as one of the powerful forces of servant leadership (Tran & Spears, 2020). Servant leaders are expected to follow the footsteps of God who never fails to work for the welfare of his people. God is acknowledged not only as one who wholesomely heals but provides the needs of his people, as in the book of the Psalms: "He will protect you from the rain and from your tears. God is all-powerful, full of love, and actively seeks your wellbeing. If you ask Him, He will provide you warmth and shelter" (Psalm 91:4). By taking an active role in promoting the mental and emotional strength of their employees, servant leaders typically inspire an exceptional level of trust and faith from others (Greenleaf, 2007; Northouse, 2019).

2.3.4 Consciousness

A servant leader should have a good sense of self-awareness. He or she should always strive to grow wiser, learning from experience. To effectively lead as a servant, the leader should be open to feedback and willing to change (Greenleaf, 2003). Aside being consistent in actions and morality and development of self, his greatest success

[1] All citations relating to the Bible were consulted in Good News Translation, Today's English Version, St. Pauls Edition (United Bible Societies, 1976).

would largely lie on his/her ability to manage emotions, anger or stress (Salas-Vallina et al., 2020). Ultimately, leading as a servant also could entail living out the principles of leadership one espouses (Greenleaf, 2007). Unlike the many people in positions of power who are thought to be blissfully ignorant of their shortcomings, the servant leaders are expected to be completely aware of their strengths, weaknesses, values, emotions and feelings (Sun, 2013). This self-awareness allows the servant leader to understand personal biases and set them aside while making decisions.

2.3.5 Forecasting

A major characteristic of a true servant leader is foresight, that is, his/her ability to stay ahead of what is happening around. A servant leader should be poised to evaluate decisions and possible consequences, considering unintended consequences (Greenleaf, 2007). In the ambit of foresight, everything is connected: the past, the present and future (Wilson, 2017). Servant leaders have an intuitive ability to predict what is likely to happen in future, based on the past and the present. This foresight enables these leaders to plan ahead. Having foresight puts the leader and the lead on a balanced ground. It offers a form of protection and safety to the leader as seen in the book of Proverbs: "Discretion will preserve you; Understanding will keep you" (Proverbs 2:11). It enhances our perception of what will occur or be produced if a certain course is followed. Thus, it increases our discernment and sharpens our judgement about which way we should go.

2.3.6 Conceptualization

An important quality of a servant leader is his/her ability to conceptualize or imagine the possibilities of the future and reconcile it with current realities (Greenleaf, 2003). This enables them to shape the present in the light of a better future. A servant leader should be able to raise and extend the others' line of vision, and provide direction and raise hope (Wilson, 2017). This ability helps the leader visualize a bright future and take the necessary steps to get there. The appropriate application of this quality is essential for his holistic growth and greater development.

2.3.7 Persuasion

Persuasion is an important characteristic of a servant leader that entails convincing others rather than coercing them through his/her positional authority (Tran & Spears, 2020). This is because, through persuasive skills, it becomes easier for a servant leader to influence the opinions and actions of others (Wilson, 2017). This quality is most useful in negotiations with business partners, customers and stakeholders. Since servant leaders are committed to the welfare of others, they use this ability only to influence others positively (Greenleaf, 2007). Servant leaders can use persuasion to build consensus and gain support from their followers (Anderson & Middleton, 2014; Dierendonck et al., 2014; Talya et al., 2019). It helps the servant leaders to negotiate on consistent principles and interests rather than positions, making everyone to feel like they have a stake in the team's success (Allen et al., 2016). This could eventually lead to an elevation of followers and promote organizational growth.

2.3.8 Stewardship

Servant leadership is deeply rooted in stewardship, in the sense that servant leaders are primarily called to be stewards. In the words of Wilson (2017), the servant leaders act as stewards for the organizations' resources. They assume complete responsibility for planning and managing all available resources for the betterment and prosperity of the organization, employees and stakeholders. Stewards are recognized as leaders when David called a great assembly that included the stewards that had oversight over everything that belonged to the king and his sons and made known to them his proposal to build a house of rest for the ark of the covenant of the Lord, as found in 1 Chronicles: "Now, David assembled at Jerusalem all the leaders of Israel: the officers of the tribes and the captains of the divisions who served the king, the captains over thousands and captains over hundreds, and the stewards over all the substance and possessions of the king and of his sons, with the officials, the valiant men, and all the mighty men of valor" (Chronicles 28:1). Given the magnitude of this responsibility, stewards need to equip themselves with wisdom so as to effectively render great service to the people. Stewards are charged with the overseeing affairs of their people and are admonished to be of good conduct always as seen in the book of Titus: "For an overseer, as God's steward, must be above reproach. He must not be arrogant or quick-tempered or a drunkard or violent or greedy for gain" (Titus 1:7). A leader with an intention to lead as a servant should also represent good sense of stewardship (Spears, 2010).

2.3.9 Commitment to People Growth

Commitment to developing the potential of those being led serves as one of the greatest strengths of servant leaders. According to Blanchard and Hodges (2003), one who strives to be a servant leader must elevate the growth and development of people from a "means" goal to an "end" goal of equal importance to the product or service mission of the organization. "Servant leadership requires a level of intimacy with the needs and aspirations of the people being led that might be beyond the level of intimacy an ego-driven leader is willing to sustain" (Blanchard & Hodges, 2003, p. 12). Leaders who employ this quality are thought to enjoy greater acceptance and would command respect from their followers (Spears, 2010). Followers become more able to realize their full capabilities when leaders nurture them, help them with their personal goals and give them control.

2.3.10 Community Building

A major strength of servant leaders lies in their ability to seek to identify some means for building community (Spears, 2010) among those who work within a given organization or institution. This, in turn, can help make those served grow as persons in the communities where they operate (Greenleaf, 2007). Unarguably, growing people is good for the individual, the organization and those the organization serves (Francis, 2015; Keith, 2014).

To always commit to the development of followers within an organization, the servant leader needs to focus on the individuals in order to identify high potential followers, understand their competencies and potential for growth, and present them with an opportunity to develop themselves through experience, support and guidance. In this regard, Greenleaf (2003, p. 8) has posited that: "all that is needed to rebuild community as a viable life form for large numbers of people, is for enough servant leaders to show the way, not by mass movements, but by each servant leader demonstrating his or her unlimited liability for a quite specific community-related group".

3 Research Design

This chapter presents a qualitative research that aims to analyse different people's voices, meanings and events regarding the concept, practice and implications of servant leadership. It is intended to help the researchers get a deeper understanding of the participants' inner experiences and to figure out how meanings are shaped through and in culture (Corbin & Strauss, 2008; Richardson, 2012).

Using an exploratory single-case study design, the study investigates the views of Church leaders and lay Christians (who are also leaders in different areas) on the concept and practice of servant leadership and its impact on the ministry of the Church and Christian community. The exploratory case study was used because it explores a phenomenon—servant leadership—in an understudied population—religious and lay Christians. A single-case study is organized around a single instance of interest and without no attention to generalizing, in order to investigate a contemporary phenomenon in depth and within its real-world context (Yin, 2013).

3.1 Research Population and Sampling Method

The target population of this study consisted of 15 participants aged between 39 and 64 years. Five among them were the clergy: three Bishops (two of whom were university lecturers) and two priests (one parish priest and one researcher); two religious women of the Catholic Church (one of whom was instrumental in the canonization of Lucia, one of the children to whom the Blessed Virgin Mary had appeared in Fátima,

Portugal); and two occupied leadership positions in the religious houses in Argentina and Portugal. The remaining eight correspond to lay people with different occupations in various organizations (one ambassador, one university lecturer, three high school teachers, one local government president, one company manager and one sports coach), working in different geographical locations (Europe, America and Africa). Table 1 summarizes some characteristics of the study participants, namely sex, age, education, occupation and region. To guarantee anonymity and confidentiality, interviewees were coded as follows: RCB_{1-3}: Clergy Bishops; RCP_{1-2}: Clergy Priests; RRW_{1-2}: Religious Women; LR_{1-7}: Lay Persons; RA_1: Atheist.

The decision to interview both clergy and lay people is considered of utmost importance because the Church, whose leadership is under study, is made up of

Table 1 Characterization of the sample

Participants	Sex	Age	Education	Occupation	Region
INTV 1 (RCB_1)	M	64	Humanities/theology	Bishop	Europe
INTV 2 (RCB_2)	M	57	Philosophy/theology	Bishop	Europe
INTV 3 (RCB_3)	M	57	Theology	Bishop	Europe
INTV 4 (RCP_1)	M	40	Theology/philosophy	Research fellow, Louvain	America, Europe
INTV 5 (RCP_2)	M	39	Theology	Parish priest	Europe
INTV 6 (RRW_1)	F	48	Medic/theology	Rev. Sister/Director	Europe, America
INTV 7 (RRW_2)	F	46	Theology	Rev. Sister/Director	Europe
INTV 8 (LR_1)	F	39	Psychology	Local government President	Europe
INTV 9 (LR_2)	M	49	Management	Company manager	Europe
INTV 10 (LR_3)	M	62	Law	Ambassador	Africa, Europe
INTV 11 (LR_4)	M	43	Philosophy/Education	Head of Department	America, Africa
INTV 12 (LR_5)	M	39	High school	Tennis coach	Europe
INTV 13 (LR_6)	F	49	Teaching degree	Teacher	Europe, Africa
INTV 14 (LR_7)	F	57	Education	Teacher/school owner	Africa, Europe
INTV 15 (RA_1)	M	55	Physical education	Teacher/coach	Europe

different people and exists for the benefit of all, serving both members and non-members. Such purpose further justifies this attempt at shedding a methodical light into the life and *raison d'être* of the Church. This was also informed by the belief that views and experiences about leadership could emerge as a result of individual characteristics and dynamics including behaviour, personality, involvement and traits (Lu et al., 2019). Therefore, interviewing different people, from within and outside the Church, would allow for a more comprehensive understanding of the phenomenon under study.

3.2 Method and Instrument for Data Collection

Data was collected using semi-structured interviews. This was considered the most effective method for data collection as the researchers aimed to explore participants views about servant leadership in the context of Church and delve into their personal experiences on the topic. As an interactive approach, it has a flexible structure since the design can be constructed and reconstructed to a greater extent (Maxwell, 2012). It is also known to elicit deeper insights into designing, administering and interpreting findings while exploring participants' behaviour, perceptions, feelings and understanding (Rahman, 2017).

Interview questions were informed by existing literature on servant leadership. The accumulated body of research on this issue was used to inspire the design of the interview guide. The questions comprising the interview guide were organized around three main themes. The first one entails the questions regarding the interviewees' views on the concept of servant leadership, the meaning ascribed to service and the characteristics associated with the servant leader. The second theme aims to get the interviewees insights on the role of the Church at the present time, and whether they see the Church as a servant leader. Additionally, they were asked about the way in which the ten characteristics developed by Spears (2010) (based on Greenleafs' writings) are implemented by the Church. The third theme addresses the future of Church from the interviewees' viewpoint. Specifically, the study participants were asked to explain how they envision the role of Church and challenged to propose ways of making their dreams come true. Moreover, they were asked to provide their opinions on the future of leadership in the Church and opportunities for improving its use. The interview guide comprises ten questions grouped into three themes as follows:

I. The concept of servant leadership and characteristics of the servant leader.

   1. What do you understand by "Servant Leadership?".
   2. In your perspective, service is?
   3. In your perspective, what has/makes a servant leader?

II. Today's Church—A servant leader?

   4. To what extent does the Church today make people grow? How?

5. Does today's Church, in her leaders, show herself to be a servant leader?
6. How does the Church implement the ten (10) characteristics (Listening, Empathy, Healing, Consciousness, Persuasion, Conceptualization, Forecasting, Stewardship, Commitment to people growth and Community Building) associated with servant leadership?

III. The Church of Tomorrow—Envisioning the future of servant leadership.
7. What is your dream for a Church of the future?
8. Concrete proposals?
9. What do you think about the future of servant leadership in the Church?
10. How can servant leadership work better in the Church than it does today?

3.3 *Data Collection Procedures and Analysis*

The empirical study comprises 15 interviews, and data was collected as follows: seven face-to-face interviews, seven online interviews and one via e-mail response. In this last case, the interviewee responded to the interview questions by e-mail. Despite a great deal of effort to get an interview, and after a brief discussion by phone about the subject of study, it was not possible to schedule an appointment to conduct the interview. Therefore, the participant agreed to answer the questions via e-mail.

The interviews lasted approximately 35–40 min. All interviews were recorded on mobile devices, and the content was transcribed word-to-word (Flick, 2002). The interviews were conducted in different languages (Portuguese or English) according to the participants' nationality (Portuguese and Nigerian). The study participants were allowed to speak in the language they felt at ease with, implying giving various points of view and ideological perspectives without the constraints of a foreign language. Transcripts in other languages were translated into English.

Due to the geographical dispersion of interviewees, it was not possible to conduct face-to-face interviews with the participants that live outside Portugal. Apart from those who live oversees, the COVID-19 pandemic was the reason for conducting interviews using remote modes (5-Skype and 2-Zoom). During face-to-face interviews, the use of masks and social distancing were observed according to the safety measures that have been put in place by governments.

Collection of data continued until saturation occurred. Data saturation is said to occur when repetition of the data emerges and no new knowledge is being obtained from additional data (Bowen, 2008). Thematic analysis of the data was used as it is among the methods that search for themes or patterns in a given set of qualitative data (Braun & Clarke, 2006). Specifically, the researchers took a deductive approach since the thematic analysis was based on some preconceived themes derived from existing literature that enabled the organization, coding and interpretation of the data. Therefore, the main themes included in the interview guide are the ones reflected in the findings section.

The keywords and sentences that emerged relevant or potentially interesting in each transcript of the interviews were coded according to the meaning ascribed to each theme. These codes allowed a comprehensive view of the main aspects, similarities and differences within the data set. This helped researchers to determine common topics in the discussion (Melchar & Bosco, 2010) and enabled them to extract meaningful passages of text to illustrate interviewees' viewpoints around the issues of the various themes.

4 Results and Discussion

The present section summarizes the main findings on each theme and the discussion in the light of the existing literature in the field of servant leadership and its practice in Modern Church.

4.1 Theme I: Concept of Servant Leadership and Characteristics of the Servant Leader

The first theme intended to get the interviewees' views and opinions on the concepts of servant leadership and service, as well as their perceptions on what makes a servant leader.

4.1.1 Interviewees' Understanding of the Concept of Servant Leadership

Interviewees provided their own definition of servant leadership, and the different aspects highlighted are presented in Table 2. Some participants explained the servant leadership concept in terms of what it is not supposed to mean. According to one of the parish priests, superiority and authoritarianism do not fit well with the servant leadership approach (RCP_2), which means that a servant leader does not force people to follow him/her. Two of the three bishops interviewed (RCB_2 and RCB_3) match the concept of servant leadership with the work of Jesus Christ, in the sense that He served as a model. According to the literature, religious people tend to make this comparison, presenting Jesus as the best example of a servant leader (Boone & Makhani, 2012).

The participants' understanding of servant leadership vividly captures the position of Greenleaf (2003) in his writings on servant leadership and also corroborates previous studies in the literature review. As mentioned before, servant leadership was discussed as (a) a value-based leadership, an approach that puts the people at the very top and the leader at the bottom, charged with serving the employees above him

Table 2 Concept of servant leadership

Servant leadership is…	Citations/Examples
A leadership style devoid of authoritarianism	"This expression 'servant leadership' thus combined describes a style of leadership in which the leader is in an attitude that is not one of superabundance, that is not one of superiority or authoritarianism. But in an attitude that he is able to see, to be, to respect and to integrate" (INTV 5 (RCP_2)
A form of leadership that is purely selfless	"I see servant leadership as a form of leadership whereby the leader does not use his position for his own benefit or to defend his own interests or as a purely personal strategy and does not do anything for the sake of himself but always for others" (INTV 13 (LR_6)
A form of leadership in which the leader understands that he or she is a servant as well	"The few who have been able live as above should be applauded because, not every leader understands that they are servants" INTV 4 (RCP_1)
Servant leadership is all about serving the others	"Servant leadership involves one being always in service, to the weakest, with time, the sick, the poor, those that society had abandoned. So, it was a leadership not to gather things for oneself, but a leadership that was in the service of the good of others. Servant leadership is an expression that will also be used by all those who throughout centuries and also today have been in charge of, or at least, those who are entrusted with pastoral responsibilities, must also recognize themselves in this very person of Jesus" INTV 2 (RCB_2)
The expression "servant leadership" is a practical word	"Servant leadership is that type of leadership which has a conscience, in which, what interests everyone must be decided by all and executed in interaction starting from the various moments of which the work agenda is made, following up on the processes up until what is decided is implemented. Our lives have meaning only to the extent that we strive to imitate Jesus Christ himself in this aspect, (who has placed before us the greatest example of servant leadership in the image of foot washing in which he shows us how he himself serves the Church) and the great saints who are models having lived out in their lives the injunctions of Jesus Christ" INTV 3 (RCB_3)

(Yukl, 2009); (b) an "other-oriented" approach to leadership which is manifested through prioritizing of follower individual needs and interests and outward reorienting of their concern for self towards concern for others within the organization and the larger community (Eva et al., 2019); (c) an effective style able to influence an organization towards achieving their goals provided the leader possesses the necessary attitudes (Boone & Makhani, 2012); and (d) an act of "leading from behind" (Athal et al., 2021).

4.1.2 Interviewees' Understanding of the Concept of Service

Table 3 presents the participants' definitions of service. The interviewees generally recognized service as the foundation of any leadership. Their opinions follow the path of definition of service found out in the literature review, where it was explained in line with Greenleaf's writings (Greenleaf, 2003) and Tarallo (2018). Service is defined as (a) the act of assuming the state of servant; (b) the desire and willingness to sacrifice self-interest in the service to others and to make a difference in their lives is the first precept of servant leadership (Barbuto & Wheeler, 2006); (c) the act of doing useful work for a person, an organization, a Church, a country or entire humanity with the view to help them meet their needs (Dierendonck et al., 2014; Eva et al., 2019; Salas-Vallina et al., 2020).

The various definitions underline specific attributes of service that result in altruistic attitudes and behaviours from leaders. Putting followers' needs first is the priority and can require self-sacrifice for the sake of others. One of the interviewees (INTV 1 (RCB_1) argues that service is the foundation of any leadership and the ultimate purpose of any organization. This means that service is the core aspect, the fundamental characteristic of servant leadership.

4.1.3 Interviewees' Understanding of What Makes a Servant Leader

When asked about what a servant leader has or is, the participants presented them as possessing an ability, a cause/mission and a commitment, and showing some particular attributes (empathy, patience, perseverance, to name a few). Table 4 presents some traits and attributes of the servant leaders through the lenses of the study participants.

According to Bennis (2002), servant leaders make a conscious choice to serve first. In their work, they place the good of the followers over their self-interests. While building strong relationships with others, they are expected to be empathic in order to feel the joy and pain of followers, as argued by a parish priest (RCP_2). In their mission, they persevere in courses of action aimed at getting the greater good of followers (RCP_1), the organization, the community and the society at large, as suggested by Tran and Spears (2020).

Table 3 Concept of service

Service is (about)…	Citations/Examples
Efforts made to help organizations achieve the purposes for which they exist	"Service is to promote the purpose of the organization. I do not conceive of any leadership, no organization no matter how simple it is that can survive if not governed from the perspective of service, no one is a bishop for self-promotion or for satisfaction; no one is president of the republic; no one is a director of a company to make a career of the company alone or of anything else just to serve himself" INTV 1 (RCB_1)
A giving in the manner of dying to give life	"Service is giving of oneself, giving also naturally with talents, time, availability, material goods, spiritual goods" INTV 3 (RCB3)
A kind of work that requires passion and commitment	"Though some people may have several reasons for serving, in service, one does not serve with the aim to get remuneration for it but in professions and other kinds of work, people can engage in them for physical financial gains they have" INTV 4 (RCP_1)
A call to live for the others	"Service is understood as the invitation to understand that the work you do, the life you live is not lived for yourself. It is lived for the others" INTV 11 (LR_4)
Making oneself available to people whenever and wherever help is needed of one	"The life of Jesus as recorded in John 3: 16 is an example and we are all called to serve whether in the Church, the religious community or the general society" INTV 7 (RRW_2)
The giving or filling someone's need	"It lends credence to the above" INTV 15 (RA_1)
The help rendered to meet the needs of others	"Such help should be selfless and would aim at the good of the others rather than the selfish end of the giver" INTV 10 (LR_3)

4.2 *Theme II: Today's Church Role and Implementation of Servant Leadership Characteristics*

This theme reflects the interviewees' views on the current role of Church, namely the impact on individuals' growth. It also captures the study participants' perceptions regarding the Church as a servant leader. Additionally, it explores how the Church implements the characteristics associated with servant leadership.

Table 4 What makes a servant leader

A servant leader has/is…	Citations/Examples
An ability to live one's life for the others	"You have to be an empathic person; therefore, you are capable of feeling… To feel the other, to feel the joys of others, to feel the pain of others. So, there it is, one making room for the other. And, finally, I don't think I have the pretension to say everything here either. But it has to be someone who lives for others, who also directs his life towards others" INTV 5 (RCP_2)
An identified cause or probably a mission	"A servant leader should actually have. So, there is that sense of mission, sense of urgency to pursue, identifying a mission, a cause to pursue and it goes with passion it goes with commitment. The servant leader has to persevere in order to face the challenges of course. The understanding of the kind or essences of mission, passion and commitment, I talk of patience and perseverance, and then with regards to those that he or she leads, we talk of humility, compassion that is empathy and a listening ear" INTV 4 (RCP_1)
Be attentive, committed… ready to lead by example	"A servant leader must be attentive, committed, reconciliatory, balanced, enthusiastic, challenging, and authentic and ready to lead by example" INTV 8 (LR_1)
A great deal of humility with a very great dose of humanity	"I expect a servant leader to be able to see in the behaviour of others, what their weak points are, their faults, to be able to help them grow" INTV 10 (LR_2)

4.2.1 Interviewees' Views of the Extent to Which the Church Contributes Positively to Individuals' Growth

Overall, the participants believe that, despite some shortcomings, the Church remains a force for good in the world in terms of supporting the people to grow. Helping individuals to grow and succeed, and making an effort to solve their problems appear in the discourses of the study participants.

Interviewee 4 (RCP1) explains the fundamental role played by the Church in education. The following quotation illustrates his point of view:

> The Church, through the establishment of Catholic schools, provides education as a source of enlightenment, civilization and development of the people. Growth cannot come about without education. In this way, the Church contributes to raise morally sound citizens.

A female religious member (Interviewee 6, RRW_1) argues that:

> The Church is the one who listens and begins to strive to solve peoples' problems, though not yet perfectly.

Another female religious member (Interviewee 7, RRW_2) puts emphasis on the spiritual dimensions of reconciliation and healing as follows:

> The Church as a channel of grace. She sanctifies, reconciles and heals through sacraments and such sacraments serves as great treasure to the Church and its faithfuls.

Interviewee 10 (LR_3), a male lay Christian, underlines the educational and spiritual role performed by the Church:

> The Church of today makes people grow spiritually and materially by performing her statutory duties. As a teacher, the Church is involved in providing spiritual guidance, fostering unity and encouraging communal or mutual integration of its members – that is fellowship with others and making people come closer to God. I believe that the Church has been consistent in this task.

Another lay participant (Interviewee 11, LR_4) highlights the community aspects of the Church and the opportunity to learn and grow based on individual experiences and those with whom they interact. The excerpt below illustrates his point of view:

> I believe that the idea of the Church existing is not only to worship, but also creating a Church community. I also think that the Church has always been involved with education, character formation, charity and so many other things that are occasions for people to encounter themselves, learn from themselves, learn from one another and grow in that regard. The Church creates occasions for growth but may be found wanting in terms of making that a guiding principle for what the Church does. So, I have a lot of reasons to think that the growth is not orderly.

Some participants are less positive and optimistic about the Church's role, thus presenting some disenchanted views on this issue. Interviewee 8 (LR_1) argues that today's Church is not making some people grow. In the words of this female lay member:

> From my perspective, today's Church is not greatly influencing people's growth, especially young people. It will be necessary to come up with strategies that capture the interest of this age group to help them grow. To do this, young people should be called to be part of the Church, and they should see themselves as part of it.

Although from a different perspective, Interviewees 6 (RRW_1) and 7 (RRW_2), both female religious members, noted that the Church has not done so well in making some people grow. Interviewee 6 (RRW_1) argues that:

> The Church does not fully accord women a rightful place in evangelism and in the liturgy. The desired position of the Church as a servant leader is rather not paramount as it is relatively lacking in this aspect.

The above quotations, with exception of the later, are in line with the argumentation of Liden et al. (2014) that servant leaders provide followers with emotional healing and positive work experiences that support their growth and success (see also, Christensen-Salem et al., 2021; Mittal & Dorfman, 2012).

4.2.2 Interviewees' Views of the Extent to Which the Church Is a Servant Leader

The general question about the current role of Church was followed by a question that specifically addressed the Church as a servant leader. The researchers wanted the participants to be more focused, by presenting their personal views on this matter, instead of giving an overall impression of the role of the Church in today's society.

According to the interviewees, the Church is threading a progressive path to show herself a servant leader through her leaders. To show the position and underlying rationale of the interviewees' views and opinions, some quotations from the interviews are provided as an illustration.

One of the bishops interviewed (Interviewee 2, RCB_2) did not hesitate in labelling the Church as a servant leader, justifying his argument as follows:

> The Church is no doubt showing herself as a servant leader. One of the ways is the de-clericalization of the Church especially since the Vatican II Council by encouraging the lay faithful to hold offices that before were the exclusive preserve of the clergy.

Another bishop (Interviewee 3, RCB_3) gives an account of an initiative that occurred during the pandemic crisis:

> The Church really shows herself servant leader in her leaders. For example, not quite long ago, the archbishop of [...] proposed to his priests during the current COVID-19 crisis to give up their one month's salary to enable the diocese to meet the increasing needs of those whose fate was more affected by the pandemic. The priests on their part were believed to have complied overwhelmingly.

One of the lay participants, who is a manager in a company (Interviewee 9, LR_2), resorts to Francis papacy and the way his words and actions are inspiring change in the Church:

> I believe that Pope Francis has shown several examples in this regard and that a good number of the members of the clergy are taking their cue from him. I think starting with the Pope himself, the way he's been passing on the message, the way he's been trying to make the changes in the organization of the Church itself, leads me to believe that we're going back a little bit more to our roots. And I recognize that effectively, 100% of all Church leaders do not imbibe this kind of attitude. But I really believe and see, I can really witness a paradigm shift and that to me is interesting.

Another lay participant (Interviewee 10, LR_3), an ambassador, cites the example of Jesus Christ, who served the followers in his days, to emphasize the nurturing role of Church in the contemporary society by providing spiritual food:

> I think finally that, the Church meets the peoples' demand like Jesus did in the Lenten period, was able to show us that Jesus Christ was not only concerned about the welfare of the apostles but also the multitude. He came and fed them three times. So, it is the same thing the Church is doing in her leadership role which concerns the growth, both spiritually and worldly, and that's what makes the Church today a servant leader.

Interviewee 12 (LR_5), who is a tennis coach, puts the emphasis on the listening skills:

> I believe strongly that the Church is becoming increasingly open and is seriously adopting one of the characteristics of good leadership which is listening.

Interviewee 13 (LR_6), however, refuted the claim that the Church, through her leaders, shows herself to be servant leader. He said:

> I appreciate what this Pope Francis, but I would not accept the high-handed manner, which characterizes the way the clergy in general treat people.

Most of the data relating to this issue follows the direction indicated by Tran and Spears (2020).

4.2.3 Implementation of the Ten Characteristics Associated with Servant Leadership

To investigate whether the ten characteristics associated with servant leadership are put in place by the Church, the study participants were asked to give their insights on the presence of the attributes developed by Spears (1998, 2010) based on Greenleaf's writings. This was intended to get evidence from servant leadership in practice. Most interviewees agreed that in varying degrees the Church implements these characteristics.

Interviewee 4 (RCP_1) underlines some critical traits of servant leadership, namely listening, empathy and healing, as follows:

> From listening, from the empathic presence, from the healing here not understood as the healing by a medical doctor, but from the perspective of inner healing, a healing of the interior, a renewal of conscience, in short, a holistic healing of the human person. We find some sectors of the Church that dedicate themselves heart and soul to all this.

The same participant (Interviewee 4, RCP_1) refers to the sacraments as avenues through which the Church implements the abovementioned characteristics of listening, empathy and healing:

> The Church has an organized mechanism for expressing compassion and showing compassion to its members and to the world. The Church is slow in punishing erring members, most times she even offers an erring member the opportunity to retrace one's steps; the sledge hammer is only applied when all persuasion and empathizing fail to bring an offender to his or her right senses.

A lay participant (Interviewee 10, LR_3) focuses his attention on stewardship:

> The Church of course, is living its life towards stewardship. Like I have mentioned earlier, they have the priests as teachers, and they post them to all areas. You can agree with me. That's to say, they have to serve, not minding all the processes.

Interviewee 13 (LR_6), a female teacher, disagrees with the overall impression of the participants on this issue. She did not agree that the Church takes the aforementioned characteristics seriously, and criticizes the Church for some kind of neglect towards the role of women:

> I do not recognize in all places of the Church. Sometimes, they have this difficulty in listening…. The young people do think that people want things different. For instance, the poor recognition given to the work of the women in the Church. And sometimes, I do not recognize this empathy in some people that live inside a very big house with everything, and they just will say 'okay, I understand your problems' but do not do nothing, more than saying that.

This theme shows that the majority of interviewees recognized that the Church is an agent of growth, that her leaders are exemplary servant leaders and that the Church implements, both in theory and in practices, the ten characteristics associated with servant leadership (Ehrhart, 2004; Fields et al., 2015). The inputs of the Interviewee 4 (RCP_1) are compelling. He submitted that though overly disparaged in some quarters as low-performing, the Church exists only because it makes people grow. In his extensive responses, he captures the essence of leadership, which he locates not just within the hierarchy but in the Church as a whole. According to him, the Church is not yet perfect in her demonstration of leadership, but what she is doing in the area of leadership on a global scale is self-evident (Kimotho, 2019; Spears, 1995, 2010). Despite this overall positive assessment of the implementation of the servant leadership dimensions by the Church, we cannot ignore the negative impression portrayed by one of the interviewees that suggests that the Church should make actions speak louder than words.

4.3 *Theme III: The Church of Tomorrow and the Future of Servant Leadership*

This last theme intended to take a tour into the future of the Church, exploring what the participants ideally expect from the Church, what they would propose to get the Church of their dreams, how they envision the future of servant leadership and how this type of leadership can be better used. This would help to project the future of a more encompassing and compassionate Church.

4.3.1 Interviewees' Dream for a Church of the Future

The study participants made known their expectations for the Church of the future with a general interest in growth.

Interviewee 1 RCB_1 envisions the future of Church (global Church and the Church in Europe) as follows:

> The Church of the future worldwide will be a greater Church. As the statistics show, and the experts who make evaluations and analyze data in this regard testify, the Church is about the religious community with more possibilities of growth. Secondly, in Europe, it will certainly be a Church in the process of quantitative decline in the coming years.

He also made a call on the Church in Europe to be the salt of the earth and the light of the world suggesting that this would certainly be possible by the influence of the laity in the bodies, groups and organizations to which they belong.

The increase of the lay faithful participation in the Church's mission also appears in the discourse of Interviewee 4 (RCP_1):

> I want to see an expansion of the role of the lay people in the Church, especially in the African Church.

The lay faithful presence in different areas and activities can help promoting the Church's mission in various ways, as shown in previous studies (Greenleaf, 2002, 2003; Ford, 1991; Nsiah, 2013; Fields, 2015; Tran & Spears, 2020).

Interviewee 3 (RCB_3) explains his vision and expectations in the following terms:

> I dream of a Church not preoccupied with itself, with internal matters, with questions that have to be dealt with, but always with the function of being transparent in the presence of the living Christ. This would be a Church rooted firmly in the tenets of God, that has fellow-feeling and respects the ecosystem reminiscent of the encyclical letter, *Laudato si,* of the Holy Father Francis on care for our common home nº 156 where he defends that human ecology is inseparable from the notion of the common good.

One of the lay participants (Interviewee 8, LR_1) claims:

> I dream of a younger Church with ceremonies more suited to the general profile of the people.

Interviewee 11 (LR_4) calls for a more inclusive space with opportunities for all the Church members, either clergy or lay persons:

> My hope for the Church of the future is that we begin to have newer and more elastic spaces or opportunities for anyone in Church leadership, whether priest or not, to just engage with the people more directly - smell how they smell, taste how they taste and so on. And perhaps, the world could happen to truly become Christian.

Although Interviewee 15 (RA_1) is an atheist, he thinks that people do not have to ever lose the vision to make us believe in and help others. When explaining his vision of the Church of the future, he says:

> There is need for a future Church with a different purpose. I think that if one day, everybody has a good socio-economic life, if everyone can eke out a living, then there will be no need for the Church. The Church does a whole lot of things that I do not understand and to which I do not agree but in all fairness, the Church's work benefits the world at this moment.

4.3.2 Proposals for the Church

The study participants were then challenged to make concrete proposals for the Church. They presented vital suggestions which they believe would help the Church to provide better leadership to the people. To support the views of the interviewees in this matter, the following quotations are provided.

Interviewee 3 (RCB_3) says:

> I propose a reform of the Church through conversion that is achievable by the prioritizing/centralizing the word of God. The reform can only happen if there is conversion. As Pope Francis said from the beginning when he spoke of the reform of the Church, he always said the reform only happens if there is conversion.

Interviewee 1 (RCB_1) briefly states:

> My own proposal is for a continuous formation of the lay people in the Church.

Interviewee 4 (RCP_1) talks specifically about leadership and the need for new models and styles of leadership in modern Church:

> I really wish that the Church would jettison the authoritarian model of leadership of the medieval ages and opt for more fashionable models of leadership that the modern world requires. A situation where the priests are the nicest class of people, dedicated to service and serving with humility, and the parishioners are not taken for granted. Such change has become necessary and overdue owing to the nature of the modern world.

Interviewee 4 (LR_1) emphasizes the role of interaction and communication among the Church members, in order to create a tight bond between those in leadership roles and the community they serve:

> Integrating activities for those who are leading the Church. The Church should promote more non-worship activities to appeal to more people. Good communication between those in charge and the people. The Church could create more work/reflection groups.

Interviewee 12 (LR_5) proposed that places of worship should be made more attractive to the worshipers. He remarked:

> It is important to make the places of worship pleasant, attractive and comfortable for more people to come in.

Interviewee 10 (LR_3) proposed that:

> The Church embraces an inter-Christian dialogue as a means to foster unity among Christians irrespective of their individual Christian denominations.

4.3.3 Future of Servant Leadership in the Church

When asked specifically about the future of servant leadership in the Church, they share a favourable impression. Although they consider that the Church still has a long way to go for her to actually be relevant in the contemporary society, they broadly believe that the future of servant leadership is promising.

According to Interviewee 3 (RCB_3):

> The Church can touch more lives and carry out her activities through using the modern means of communication. Also, if our leaders are living the life of service, it will have a great impact on the lives of many families.

Interviewee 5 (RCP_2) links the future of servant leadership to the exemplary leadership figure of Pope Francis:

> There is a future for servant leadership if the Church leaders imitate Pope Francis' way of leadership. The present Pope has given a great encouragement and a great example to all the leaders of the Church, to all believers and non-believers alike.

In the same vein, the qualities of Pope Francis are presented as a model for other leaders in the words of Interviewee 9 (LR_2):

> Servant leadership has a future in the Church as the leaders are becoming closer and nicer to the people. The present Pope is a powerful example of a servant leader. I believe that a leader who is willing to serve has a greater positive influence on the people under his care.

Interviewee 10 (RL_3) brings to the discussion other religious leaders of the Church, such as the former Pope John Paul II:

> The future of servant leadership in the Church is clear. It has been well led by the former Pope, Pope John Paul II; the present Pope; and the other leaders. So, for me, there is a bright future for servant leadership in the Church if only, on one condition: if the leaders will not shy away from whatever the circumstance and preach the suffering Christ; live by example; be able to say to the parishioners, look, this is wrong or right. If not, the Catholic Church would not have been able to correct itself. That's one of the duties of the Church.

Interviewee 8 (LR_1) gives priority to the core concept of service:

> The Church will advance further if her leadership is servant-oriented. In this way, it becomes conciliatory and open to all the fringes of the population.

4.3.4 Making Servant Leadership Work Better in the Church Than It Does Today

When asked about the ways the servant leadership could work better in the Church, interviewees raised critical issues that would help reposition the Church when addressed.

One is for the Church to be continually asking herself if she is actually being a servant to the people and trying not to straw from the right track.

In this regard, Interviewee 1 (RCB_1) is of the opinion that:

> The Church should preach more and channel its work towards conversion and should be able to convert in service. The Church needs to imbibe the attitude of conversion, and to answer the question, how can the Church function better in this line of servant leadership? By continually asking herself whether she is actually being servant.

It was also emphasized that Church leaders should learn how to computerize their system as much as possible in order to communicate more efficiently with their members, as explained by Interviewee 3 (RCB_3):

> There is one concrete thing that comes into focus now in this time that we are living that is, in fact, learning how to use the computerized means of communication. The Church should strive to make that something habitual.

Some participants argue that the Church could do better if the leaders are free from political and social leadership and other forms of worldly influences. This can be found in the subtleties of Interviewee 2 (RCB_2) discourse:

> The Church should maximise the use of mass communication media for information dissemination, and the leaders should not allow themselves and the Church to be negatively influenced by the changing world.

Another aspect is that, in the future Church, servant leadership can work better if people, especially leaders, are all able to convert daily to the message of the Gospel. This means putting words into action.

Interviewee 5 (RCP_2) presents this argumentation as follows:

> Servant leadership can work better especially when every Christian takes cognizance that in a certain way, in the Church all the baptized are leaders and therefore no one can exclude oneself from this mission of being a leader, and therefore being a leader in this perspective of Jesus as a typical servant leader. But now it is true that to be more like Jesus we constantly need to convert to Jesus. And this permanent conversion is a requirement of the Gospel, so we cannot forget that. So, servant leadership can work better in the Church if we are all able to convert daily to the Gospel.

Also, there should be readjustment or upward movement in the manner things are done in the Church to accommodate the modern Christians, as stated by Interviewee 4 (RCP_1):

> If the Catholic Church or the whole Church wants to really readjust, adjust to turn up in response to the demand of the modern world, that's not bad. Whether we are responding or reacting, the most important thing is that there is something, there is a movement, upward movement.

Servant leadership can work better in the Church when it can create a space for balance, meeting and acceptance of various points of view—argues Interviewee 8 (LR_1). In so doing, it would be possible for the Church to attract the attention of a large number of people. Therefore, the kind of leadership that the Church has to follow is a shared leadership system—a leadership that not only leads people, convinces them and leads them to behave differently, but also shares that leadership with others, letting them know that they can as well be leaders themselves. This is in line with the commitment feature of servant leadership, in the sense of an inclusive decision-making process that encourages followers to get involved for personal and community growth (Francis, 2013; Spears, 2010).

A lay participant (Interviewee 9, LR_2) calls for the inclusion of leadership courses in the training and education programmes of religious people:

> The Church should strive to introduce more disciplines/courses in the seminaries for gone are the days when a priest can afford to be ignorant. Let me ask you a concrete question. Do you have any leadership course in the seminary curriculum?" [to which the interviewer answered in the affirmative], alluding that that might not have been the case in the past.

This concern also appears in the interview of a lay participant (Interviewee 13, LR_6):

> And there will be the servant leadership espoused by the Church and proposed to the society at large as a genuine style of leadership. In the olden days, people went to the seminary to escape from the harsh realities of life. Today the priest lives with all kinds of demands on his person that make an all-encompassing formation for him indispensable.

As remarked by Interviewee 14 (LR_7), the servant leadership would work better in the Church if the leaders tried more to improve the lives of the followers. This could be in the form of mentoring especially for the youth.

Interviewee 10 (LR_3) argues that Church leaders, and especially the lay leaders in positions of leadership, should demonstrate service by the way they live as their lives serve as examples for their followers:

> If as a parishioner, I am given a position in the Church, I will live by example. By the way I carry myself and participate in the Church, people will see me as a Christ-model leader to follow. So, by leaving for mass early; participating in the work that I am given; diligently carrying out the duties and showing the people that this is the way, the tendency is that what we have been speaking about since morning would have already been realized and I am happy to teach this. And I think that this is the way forward.

Evidence from the interviews suggests that though the time has come and gone when the Church and her pastoral and preaching activities dictated the mode of life and greatly influenced how people thought and lived, the Church has continued to be a forerunner in the defence of the human person through its members, both religious leaders and lay persons.

The clergy, most of whom are versed in both sacred and secular sciences, seemed more consistent and cautious in their analysis about the past, the present and the future of the Church. The lay people without throwing caution to the winds engaged constructively in the interview in the hope that the outcome will be productive and will add value to the way things are done by the Church and especially the way the laity is perceived. They brought in their professional experiences as ranking office holders and business managers which they believe can help the Church auto-evaluate the leadership style (Ingram, 2016; Kimotho, 2019; Spears, 2010).

5 Conclusion

The chapter presented the viewpoints of religious leaders and clergy, as well as lay persons, on the concepts of servant leadership and the characteristics of servant leaders.

From the results of our analyses, it became clear that one is only a leader to the extent that one serves. Projecting servant leadership firstly as an up-coming leadership style and secondly as that style that influences and impacts the most in people's lives in general and the Church in particular, the study succinctly addressed the means a leader adopts in order to succeed in serving others. These means would be easily identifiable among the characteristics associated with servant leadership, namely listening, empathy, healing, consciousness, persuasion, conceptualization, forecasting, stewardship, commitment to people's growth and community building. The Church is considered a servant leader even as it continues to struggle to successfully implement the characteristics associated with servant leadership. The application of the ten characteristics of servant leadership by leaders of institutions and organizations will tendentially produce a spiralling effect.

The study evokes worthwhile strides in the study of servant leadership and demonstrated how servant leadership is antecedent to any worthwhile pastoral and social ministry in the Church. Although the Church is not perfect, she is awake to her responsibility of being a moral conscience for humanity. This became clear in the voices of the study participants, both the religious leaders and the lay persons interviewed. In practical terms, the Church builds and runs hospitals, orphanages and schools at all levels. The Church maintains a presence, from cradle-to-grave, in every human situation in order to accompany humanity in the chances and changes of life. The Church does not lay any claims to perfection in doing this because she is in a continuous process of becoming herself.

The ultimate aim of servant leadership in the modern Church is to make people better citizens regardless of their positions, roles and activities. This work will hopefully alert the readers of the importance of imbibing Jesus-like attitudes in running their businesses. Servant leadership shows that the way to lead is to serve, to be humble, to be patient and to be empathetic with the led. And as is discernible from the many unsuccessful and unimplementable policies that emanate from many world leaders with little or no ethical consideration, this work serves to point out not just an alternative to unethical leadership, but a time-tested and style of leadership whose time has come. More studies are needed to investigate the existence and practice of servant leadership in other contexts and organizations.

References

Agli, O., Bailly, N., & Ferrand, C. (2015). Spirituality and religion in older adults with dementia: A systematic review. *International Psychogeriatrics, 27*(7), 15–725.

Agyei, T. B. (2019). *Assessing the role of lay leadership in the Church of Pentecost, Parkoso local.* Doctoral dissertation, Department of Theology, Faculty of Humanities, Christian Service University College, Ghana.

Aline, I., & Ramkumar, S. (2018). Leaders are not born, they are made. *International Journal of Applied Research, 4*(5), 94–96.

Allen, G. P., Moore, W. M., Moser, L. R., Neill, K. K., Sambamoorthi, U., & Bell, H. S. (2016). The role of servant leadership and transformational leadership in academic pharmacy. *American Journal of Pharmaceutical Education, 80*(7).

Alm, K., & Guttormsen, D. S. (2021). Enabling the voices of marginalized groups of people in theoretical business ethics research. *Journal of Business Ethics*, 1–18.

Anderson, G. L., & Middleton, E. B. (2014). LeaderPAR: A participatory action research framework for school and community leadership. In *International handbook of educational leadership and social (in) justice* (pp. 275–287). Springer.

Armstrong, M. (2006). *A handbook of human resource management practice.* Kogan Page.

Athal, J. L., Gowreesunkar, V., & Betchoo, N. (2021). *Servant leadership, the art of leading from behind: An exploratory study.* Academic Doctoral presentation. Center for Management Studies Jain Bangalore, India. https://www.academia.edu/s/a0c34a9dfd. Accessed 21 March 2021.

Aubrey, M. (1998). The importance of mission: Nine reasons why a mission is important. In M. Aubrey (Ed.), *Developing a dynamic mission for your ministry: Finding direction and making an impact as a Church leader* (pp. 13–27). Kegrel Publications.

Baker, M. (2014). Leadership styles and a changing world. In LEADERSHIP PSYCH 485. The Pennsylvania State University (PennState). https://sites.psu.edu/leadership/2014/06/29/leadership-styles-and-a-changing-world/. Accessed 13 May 2020.

Barbuto, J. E., & Wheeler, W. D. (2006). Scale development and construct clarification of servant leadership. *Group and Organizational Management, 31*(3), 300–326.

Bennis, W. (2002). Leaders in transition. *Executive Excellence, 19*(4), 13–13.

Berson, Y., & Avolio, B. J. (2004). Transformational leadership and the dissemination of organizational goals: A case study of a telecommunication firm. *The Leadership Quarterly, 15*(5), 625–646.

Blanchard, K., & Hodges, P. (2003). Self-serving leaders vs. servant leaders. In K. Blanchard & P. Hodges (Eds.), *The servant leader: Transforming your heart, head, hands and habits* (pp. 17–18). Thomas Nelson Inc.

Boone, L. W., & Makhani, S. (2012). Five necessary attitudes of a servant leader. *Review of Business, 33*(1).

Bowen, G. A. (2008). Naturalistic inquiry and the saturation concept: A research note. *Qualitative Research, 8*(1), 137–152.

Braun, V., & Clarke, V. (2006). Using thematic analysis in psychology. *Qualitative Research in Psychology, 3*(2), 77–101.

Burkus, D. (2010) Servant leadership theory. https://davidburkus.com/2010/04/servant-leadership-theory/. Accessed 29 April 2020.

Canavesi, A., & Minelli, E. (2021). Servant leadership: A systematic literature review and network analysis. *Employee Responsibilities and Rights Journal*, 1–23.

Chiniara, M., & Bentein, K. (2016). Linking servant leadership to individual performance: Differentiating the mediating role of autonomy, competence and relatedness need satisfaction. *The Leadership Quarterly, 27*(1), 124–141.

Christensen-Salem, A., Marco, T. F., Zanini, M. T. F., Walumbwa, F. O., Parente, R., Peat, D. M., & Perrmann-Graham, J. (2021). Communal solidarity in extreme environments: The role of servant leadership and social resources in building serving culture and service performance. *Journal of Business Research, 135*, 829–839.

Corbin, J., & Strauss, A. (2008). *Basics of qualitative research.* Sage.

DeRue, D., & Ashford, S. (2010). Who will lead and who will follow? A social process of leadership identity construction in organizations. *Academy of Management Review, 35*(4), 627–647.

Dierendonck, D. V., Stam, D., Boersma, P., Windt, N. D., & Alkema, J. (2014). Same difference? Exploring the differential mechanisms linking servant leadership and transformational leadership to follower outcomes. *The Leadership Quarterly, 25*(3), 544–562.

Ehrhart, M. (2004). Leadership and procedural justice climate as antecedents of unit-level organizational citizenship behaviour. *Personnel Psychology, 57*(1), 61–94.

Embree, R. (2016). *11 Characteristics the Church needs in leaders today.* Opinion in good faith media. https://goodfaithmedia.org/11-characteristics-the-Church-needs-in-leaders-today-cms-23205/. Accessed 03 February 2020.

Eva, N., Robin, M., Sendjaya, S., Van Dierendonck, D., & Liden, R. C. (2019). Servant leadership: A systematic review and call for future research. *The Leadership Quarterly, 30*(1), 111–132.

Fairholm, R. M. (2015). *Defining leadership: A review of past, present, and future ideas.* Center for Excellence in Municipal Management's Research Monograph Series MS02-02, 1–45.

Fields, J. W., Thompson, K. C., & Hawkins, J. R. (2015). Servant leadership: Teaching the helping professional. *Journal of Leadership Education, 14*(4), 1–14.

Flick, U. (2002). Qualitative research-state of the art. *Social Science Information, 41*(1), 5–24.

Ford, L. (1991). *Transforming leadership: Jesus' way of creating vision, shaping values, and empowering change.* InterVarsity Press.

Francis, P. (2013). *Evangelii gaudium.* United States Conference of Bishops.

Francis, P. (2015). Laudato si. Vatican City. Vatican Press, May, 24, w2.

Fulenwider, R. (2004). Servant leaders. In R. Fulenwider (Ed.), *The servant-driven Church: Releasing every member for ministry* (pp. 43–65). College Press Publishing Company.

Gardner, J. (1990). *On leadership*. Free Press.
Gesme, D., & Wiseman, M. (2010). How to implement change in practice. *Journal of Oncology Practice, 6*(5), 257–259.
Greenleaf, R. K. (1970). *The servant as leader*. Greenleaf Publishing Center.
Greenleaf, R. K. (2002). *Servant leadership: A journey into the nature of legitimate power and greatness*. Paulist Press.
Greenleaf, R. K. (2003). *The servant leader within: A transformative path*. Paulist Press.
Greenleaf, R. K. (2007). The servant as leader. In W. C. Zimmerli, M. Holzinger, & K. Richter (Eds.), *Corporate ethics and corporate governance* (pp. 79–85). Springer.
Hale, J. R., & Fields, D. L. (2007). Exploring servant leadership across cultures: A study of followers in Ghana and the USA. *Leadership, 3*(4), 397–417.
Ingram, O. C. (2016). Servant leadership as a leadership model. *Journal of Management Science and Business Intelligence, 1*(1), 21–22.
John Paul, I. I. (1991). Encyclical letter. *Centesimus Annus, 36*, 838–840.
Kahler. (2000). Mission and evangelism: The emergence of the theological challenge. In D. L. Guder (Ed.), *The continuing conversion of the Church* (pp. 3–27). William B. Eerdmans Publishing Company.
Keith, K. M. (2014). Servant leadership: The perfect fit for family business. *Today's Manager, 3*, 1–5.
Khan, Z. A., Nawaz, A., & Khan, I. (2016). Leadership theories and styles: A literature review. *Journal of Resources Development and Management, 16*(1), 1–7.
Kimotho, S. G. (2019). Is servant leadership a 'Christian theory'? A critical examination of Greenleaf's servant leadership concept. *International Journal of Research in Humanities and Social Studies, 6*(3), 71–78.
Lamm, K. (2021). *10 Leadership mistakes in transitioning a Church's worship*. https://www.renewingworshipnc.org/10-leadership-mistakes-in-transitioning-a-churchs-worship/. Accessed 22 February 2021.
Leibensperger, M. S. (2015). Servant leadership and follower attributes. In *LEADERSHIP PSYCH 485*. The Pennsylvania State University (PennState). https://sites.psu.edu/leadership/2015/11/08/servant-leadership-and-follower-attributes/. Accessed 14 July 2020.
Liden, R. C., Wayne, S. J., Liao, C., & Meuser, J. D. (2014). Servant leadership and serving culture: Influence on individual and unit performance. *Academy of Management Journal, 57*(5), 1434–1452.
Liew, M. (2017). *Everyone is a little egocentric*. Psychology & Mental Health. https://www.learning-mind.com/egocentric-narcissistic/. Accessed 11 August 2020.
Ling, Y., Simsek, Z., Lubatkin, M. H., & Veiga, J. F. (2008). The impact of transformational CEOs on the performance of small to medium-sized firms: Does organizational context matter? *Journal of Applied Psychology, 93*(4), 923–934.
Lobl, T. (2018). Growing beyond 'us and them'. Editorial, Christian science sentinel. May 14, 2018, Issue, 1–3. https://sentinel.christianscience.com/issues/2018/5/120-20/growing-beyond-us-and-them Accessed 21 August 2020.
Logan, T. C. (2011). *Growing beyond an egocentric worldview*. https://www.academia.edu/4461646/Growing_Beyond_an_Egocentric_Worldview. Accessed 29 December 2020.
Lu, J., Zhang, Z., & Jia, M. (2019). Does servant leadership affect employees' emotional labor? A social information-processing perspective. *Journal of Business Ethics, 159*(2), 507–518.
Malone, J., & Dadswell, A. (2018). The role of religion, spirituality and/or belief in positive ageing for older adults. *Geriatrics, 3*(2), 28.
Maxwell, J. A. (2012). *Qualitative research design: An interactive approach* (pp. 41–52). Sage.
McMinn, T. F. (2001). The conceptualization and perception of biblical servant leadership in the Southern Baptist convention. *Digital Dissertations*, 3007038.
Melchar, D. E., & Bosco, S. M. (2010). Achieving high organization performance through servant leadership. *The Journal of Business Inquiry, 9*(1), 74–88.

Mittal, R., & Dorfman, P. W. (2012). Servant leadership across cultures. *Journal of World Business, 47*(4), 555–570.
Northouse, P. G. (2019). *Leadership: Theory and practice* (8th ed., pp. 347–390). Sage.
Nsiah, G. K. B. (2013). Leading as Jesus led: Christ models of leadership. *Open Journal of Leadership, 2*(4), 103–105.
Parris, D. L., & Peachey, J. W. (2013). A systematic literature review of servant leadership theory in organizational contexts. *Journal of Business Ethics, 113*(3), 377–393.
Rachmawatia, A. W., & Lantub, D. C. (2014). Servant leadership theory development & measurement. *Procedia - Social and Behavioral Sciences, 115*, 387–393.
Rahman, S. (2017). The advantages and disadvantages of using qualitative and quantitative approaches and methods in language "Testing and assessment" research: A literature review. *Journal of Education and Learning, 6*(1), 102–112.
Reddy, A. V., & Kamesh, A. V. S. (2016). Integrating servant leadership and ethical leadership. In M. Chatterji & L. Zsolnai (Eds.), *Ethical leadership* (pp. 107–124). Palgrave Macmillan.
Rego, A., & Cunha, M. P. (2008). Workplace spirituality and organizational commitment: An empirical study. *Journal of Organizational Change Management, 21*(1), 53–75.
Rego, A., Cunha, M. P., Gomes, J. F. S., Cunha, R. C., Cabral-Cardoso, C., & Marques, C. A. (2015). *Manual de Gestão de Pessoas e do Capital Humano* (3rd ed.). Edições Sílabo.
Richardson, A. J. (2012). Paradigms, theory and management accounting practice: A comment on Parker (forthcoming) "Qualitative management accounting research: Assessing deliverables and relevance." *Critical Perspectives on Accounting, 23*(1), 83–88.
Rost, J. (1991). *Leadership for the twenty first century*. Praeger.
Ruiz-Prada, M., Fernández-Salinero, S., García-Ael, C., & Topa, G. (2021). Occupational stress and catholic priests: A scoping review of the literature. *Journal of Religion and Health, 60*(6), 3807–3870.
Russell, R. F. (2001). The role of values in servant leadership. *The Leadership & Organization Development Journal, 22*(2), 76–83.
Russell, R. F., & Stone, A. G. (2002). A review of servant leadership attributes: Developing a practical model. *Leadership & Organization Development Journal, 23*(3), 145–157.
Salas-Vallina, A., Simone, C., & Fernández-Guerrero, R. (2020). The human side of leadership: Inspirational leadership effects on follower characteristics and happiness at work. *Journal of Business Research, 107*, 162–171.
Samuel, S., & Sugden, C. (1987). *Transformation: The Church in response to human need*. Wm. B. Eerdmans Publ. Co. and Regnum Books.
Savel, R. H., & Munro, C. L. (2017). Servant leadership: The primacy of service. *American Journal of Critical Care, 26*(2), 97–99.
Sendjaya, S., Sarros, J. C., & Santora, J. C. (2008). Defining and measuring servant leadership behaviour in organizations. *Journal of Management Studies, 45*(2), 402–424.
Setley, D. M., & Gautsch, D. (2015). Leadership and the Church: The impact of shifting leadership constructs. *International Journal of Business and Social Research, 05*(12), 15–25.
Spears, L. (1995). Servant-leadership and the Greenleaf legacy. In L. Spears (Ed.), *Reflections on leadership: How Robert K. Greenleaf's theory of servant-leadership influenced today's top management thinkers* (pp. 1–14). Willey.
Spears, L. C. (Ed.). (1998). *Insights on leadership: Service, stewardship, spirit, and servant-leadership*. Wiley.
Spears, L. C. (2010). Character and servant leadership: Ten characteristics of effective, caring leaders. *The Journal of Virtues & Leadership, 1*(1), 25–30.
Stone, A., Russell, R. F., & Patterson, K. (2003). Transformational versus servant leadership: A difference in leader focus. *Leadership & Organization Development Journal, 25*(4), 349–361.
Sun, P. Y. T. (2013). The servant identity: Influences on the cognition and behavior of servant leaders. *The Leadership Quarterly, 24*(4), 544–557.
Swaroop, K. R., & Prasad, N. G. A. (2013). Are leaders born or made? *Asia Pacific Journal of Marketing & Management Review, 2*(8), 35–40.

Talya, N., Bauera, T. N., Perrot, S., Robert, C., Liden, R. C., & Erdogan, B. (2019). Understanding the consequences of newcomer proactive behaviors: The moderating contextual role of servant leadership. *Journal of Vocational Behavior, 112*, 356–368.

Tarallo, M. (2018). *The art of servant leadership*. SHRM. Org.

Taylor, R. J., Ellison, C. G., Chatters, L. M., Levin, J. S., & Lincoln, K. D. (2000). Mental health services in faith communities: The role of clergy in black churches. *Social Work, 45*(1), 73–87.

Tran, D. Q., & Spears, L. C. (2020). Servant-leadership and community: Humanistic perspectives from Pope John XXIII and Robert K Greenleaf. *Humanistic Management Journal, 5*(1), 117–131.

Walumbwa, F. O., Avolio, B., Gardner, W. L., Wernsing, T. S., & Peterson, S. J. (2008). Authentic leadership: Development and validation of a theory-based measure. *Journal of Management, 34*(1), 89–126.

Webster, L. (2003). The call to leadership is a call to service, building both Church and self. *Understanding our Church: A treasury of Arkansas writers discussing the catholic faith*. https://www.dolr.org/article/call-leadership-call-service-building-both-church-and-self Accessed 21 August 2020.

White, P., & Acheampong, B. O. (2017). Planning and management in the Missional agenda of the 21st Century Church: A study of Lighthouse Chapel International. *Verbum Et Ecclesia, 38*(1), 1–8.

Wilson, R. (2017). 12 Characteristics & behaviours of an authentic servant leader. https://growingorganisations.com/12-characteristics-behaviours-authentic-servant-leader/. Accessed 29 September 2020.

Yin, K. R. (2013). Validity and generalization in future case study evaluations. *Evaluation, 19*(3), 321–332.

Yukl, G. (2009). Leading organizational learning: Reflections on theory and research. *The Leadership Quarterly, 20*(1), 49–53.

Zehndorfer, E. (2013). *Leadership: A critical introduction*. Routledge.

Zisa, J. (2013). Listen to serve: Servant leadership and the practice of effective listening.

Homosexual Collaborators' Perception About Discrimination by the Organization and/or Peers: A Case Study

Carolina Machado and Liliana Isabel Mendes Vilarinho

Abstract The present study focuses on the perception of homosexual individuals, in the labor context, either by heads or pairs, about discrimination. In light of concepts such as coming out, homophobia vs. heteronormativity and management of organizational diversity, this exploratory case study intends to align some concepts, doubts and hypotheses that have emerged throughout the literature review in order to foster future work. This chapter looks to analyze common points among the interviewed people, confronting, at the same time, with the literature. So, we have developed two interviews with two homosexual individuals who answered questions related to their sexual orientations' perceptions, both at individual level and in society in general as well as in the workplace context (functions, managers and peers). This study looks to identify the employee's perception about sexual discrimination by the organization where he/she works, and the peers and whether the individual should reveal or omit his / her sexual orientation. The analysis of the interviews suggested some facts different from what was read. For example, the interviewees don't perceive discrimination and reveal that less and less this difference is felt in the work context. In short, it is also intended to "open doors," sensitizing and showing that the management of diversity in a labor context is of great importance, not only for the individual in his individuality, but also in the social / labor context, as well as in the performance of their functions and, consequently, the dynamics of the organization.

Keywords Coming out · Homophobia · Heteronormativity and management of organizational diversity

C. Machado (✉)
School of Economics and Management, Interdisciplinary Centre of Social Sciences (CICS.NOVA.UMinho), University of Minho, Braga, Portugal
e-mail: carolina@eeg.uminho.pt

L. I. M. Vilarinho
School of Economics and Management, University of Minho, Braga, Portugal

C. Machado (ed.), *Challenges and Trends in Organizational Management and Industry*, Management and Industrial Engineering,
https://doi.org/10.1007/978-3-030-98048-1_3

1 Introduction

Rohm et al. (2012) reinforce that in organizations, one of the most marginalized groups is that of homosexuals. According to Irigaray and Freitas (2013), referring to Alves and Galeão-Silva (2004), the work environment in organizations is increasingly becoming a heterogeneous space—a space where individuals of different genders, ethnicities, religions and sexual orientations coexist. When we look to the existent literature, we found several approaches to sex discrimination, as well as other types of discrimination, suffered at work. However, and looking to the Portuguese context, we expected to find more articles and researches concerning this topic. Indeed, we found that research in this area is still scarce and dispersed among several approaches more biased toward discrimination in general and based on theoretical assumptions. This is why we decided to explore the topic with regard to the perception of homosexual employees about discrimination by the organization and/or peers.

Several questions arose throughout the literature review, but the search for an answer was always guided by the concealment of certain details, as is understandable by the sensitivity that subjects like homosexuality arouse.

In a country with a society as heterogeneous as the Portuguese, it is essential that the study of diversity achieves the due relevance so that organizations and society in general, free from prejudice, facilitate access to the labor market for all citizens equally. With this, the practice of diversity management becomes relevant in order to attract new talent, promoting and developing competence and, consequently, increasing the efficiency and productivity of organizations (Rohm et al., 2012).

In order to develop this chapter, after reviewing the literature, an overview of the main concepts that will be integrated into the analysis of the subject will be made. As a way to complete this analysis, it will be carried out an interview with two homosexuals where the perception they have about the acceptance of their sexual orientation, either in the organization where they work and by co-workers, will be reported. After analyzing these data, a confrontation between the literature and the data collected in the interviews will be made.

The main aim of this chapter is to show that diversity exists and is necessary, and that it should be managed in order to develop individualities and promote collectiveness in organizations, develop and retain talent and, consequently, promote efficiency, effectiveness and productivity of organizations.

The chapter is structured, in a first moment, with the definition of concepts, followed by the methodology used in this exploratory study and, soon after, the analysis of the interviews. It ends with the final considerations.

2 Exploration and Definition of Concepts

To explore this topic, some concepts should be explained, such as the concept of coming out, homophobia vs. heteronormativity and management of organizational diversity.

2.1 *Sexual Identity—Coming Out*

The process of "coming out" which we can translate into Portuguese slang as "coming out of the closet" is a process of formation of a homosexual identity.

Several authors have analyzed this process and defined it as:

> A complex process of interpersonal transformations, which goes from adolescence to adulthood, and which leads to a set of events recognizing the individual's sexual orientation (Haneley-Hackenbruck 1989, referred by Wilson, 1999, p. 560);
>
> Being a developmental process in which sexual preferences are recognized by homosexual individuals and they choose to communicate and make this knowledge integrated into their personal and social lives (Monteflores and Schultz 1978, referred by Zera, 1992).

With this, it can be said that despite the definitions of this process—coming out—be based on the individual's personal identity, they are imprinted and integrated into a wider social dimension.

In line with this idea, we see today that the average age of coming out is increasingly precocious and, not only, in adulthood (Frazão & Rosário, 2008). In other words, the awareness of sexual identity is increasingly a challenge for individuals in adolescent age.

Nevertheless, it is a fact that this process is complex and there are some models to explain it.

In their article, Frazão and Rosário (2008) make a compendium of several authors about the most classic coming out models:

*–**Cass Model (1979)*** Proposes a stadium system (6 stadiums) that describes the coming out, namely:

1. *Confusion of identity*—Personalization of information about one's sexuality: "Will I be a homosexual?". First, the recognition of homosexual thoughts/behaviors as being unacceptable and, later, the search for information to redefine the meaning of these thoughts/behaviors.
2. *Comparison of identity*—Acceptance of the individual's possibility to be homosexual: "I might even be homosexual." Recognition of differences in relation to others and tendency to socially isolate themselves. However, the individual even accepting his homosexual identity inhibits his/her behavior as such.
3. *Identity Tolerance*—Recognition of the sexual, emotional and social needs of being a homosexual: "I'm probably even a homosexual". The individual begins

to build a sense of belonging to a community by looking for other homosexuals in groups or bars, initiating a greater exploration of sexuality.

4. *Acceptance of identity*—Acceptance of oneself as a homosexual and, consequently, contact with the homosexual culture is intensified: "Who am I? Where do I belong?" The indignation of society's anti-gay or homophobic opinions grows.
5. *Pride of identity*—There is a greater interaction with the homosexual culture at the expense of the heterosexual one: "I'm proud to be gay." The individual gives greater importance to who is homosexual beginning to have a confrontation with heterosexual culture. There is more corporate behavior and disclosure of sexual identity to friends, family and colleagues begins.
6. *Synthesis of identity*—Sexual identity, although important, is no longer essential in the relationship with others and there is recognition of heterosexual people who are a source of support: "I know who I am. Being gay is one of the many things I am."

–***Coleman's model (1982)*** - Characterizes the coming out process in five stages:

1. *Pre-coming out* —The child feels different from their peers. However, for many years, the child incorporates dominant values of society, which considers homosexuality wrong, which leads to feeling different, alienated and alone. The individual protects himself because there is a clear perception that when feelings are directed toward people of the same sex, they are the target of rejection and ridicule.
2. *Coming Out*—Recognition of homosexual feelings with the individual having the perception and understanding of what it means to be homosexual and preparing to share with others (friends, colleagues or family members) revealing their sexual orientation.
3. *Exploration*—Development of socialization skills with people with the same sexual interests, experimenting with a new sexual identity and the sense of attractiveness and sexual competence. However, it is essential to understand that sexual conquest is not the basis of an individual's self-esteem.
4. *First relationships*—There are needs in terms of intimacy and, consequently, the desire to have a stable and committed relationship. At this stage, the individual learns to function in a homosexual relationship in a society where heterosexual relationships are the norm.
5. *Integration*—Public Identity and Private Identity Are United in a Single Self-Image.

In short, the models by Cass and Coleman summarize some information on how the identity process of a homosexual individual permeates the achievement of a global identity that incorporates their sexual experience.

Nevertheless, the more classic ones, still in the same article, Frazão and Rosário (2008), refer to other authors who tried to build more inclusive and integrative models, such as Ritter and Tendrup (2002, referred by Pachankis & Goldfried, 2004).

These authors proposed a synthesizing model of all the coming out models already existing in the literature, from the 1970s to the present day, identifying three major phases common to all of them:

1. *Sensitization*—The individual feels that he/she is different and marginalized in relation to same-sex peers, and feels that the gender roles traditionally stipulated by society do not fit him/her. He/she develops defense strategies to overcome this non-acceptance, such as taking homophobic positions, immersing himself/herself in a heterosexual identity or thinking that it is just a phase, among other strategies.
2. *Tolerance*—The homosexual individual does not reveal his/her sexual identity by engaging in a double life. It is common for homosexual individuals to have heterosexual relationships with friends and family and, at the same time, to have homosexual involvements to suppress their sexual, emotional and social needs. The individual maintains this double life until he/she endures the dissonance after which he/she moves to the stage of being proud of being a homosexual and seeks intimate relationships with people of the same sex more assiduously.
3. *Integration*—the Individual Will Integrate Their Homosexual Identity into His/her Overall Self.

The efforts of these authors to create more integrative or more traditional models have been the target of several criticisms as these models are too normative and rigid, not doing justice to the reality of the diversity of life paths of homosexual individuals (Frazão & Rosário, 2008).

Many homosexuals decide to reveal their sexual identity to closest people because they feel they cannot keep that identity a secret and sharing will begin the literal experience of the expression "coming out", creating in the individual a feeling of freedom and honesty toward themselves and in interpersonal relationships that he/she establishes.

According to these authors, this process of coming out can be liberating, but at the same time these individuals may face emotional and behavioral problems. Effectively, problems associated with phobias, depression, suicide attempts, running away from home, substance abuse, sexual promiscuity and others can arise here.

2.2 *Homophobia vs. Heteronormativity*

In the discussion of diversity management in the organization, with regard to sexual orientation, we must always think and understand the concepts of homophobia and heteronormativity, since, according to Irigaray (2006), intolerance to different sexual orientations results in compromising the dignity of these individuals, consequently increasing their stress level.

In general, according to Dinis (2011), homophobia is defined as prejudice and discrimination against homosexual people. Costa and Nardi (2015) report that the concept of homophobia has been widely used when the subject is violence and

Table 1 Prejudice against sexual diversity

Prejudice against sexual diversity		
Beliefs ('Homosexuals should not be teachers')	Affections ('I feel angry at homosexuals')	Behaviors ('I don't select gay candidates')

Source Adapted from Costa and Nardi (2015)

discrimination against individuals who have a different sexual orientation than heterosexual (Table 1)

According to Irigaray and Freitas (2013), homophobia is the unconscious acceptance and adoption of negative feelings, ideas and attitudes against homosexuals.

However, when there is a belief that heterosexual orientation is superior and, at the same time, homosexual individuals are deliberately excluded, for instance, from public and organizational policies, and social events, we are talking about heteronormativity.

To these authors (referring to Gutkoski, 2006), heteronormativity stigmatizes, denies and denigrates any and all non-heterosexual cultural manifestations, from performing civil and religious weddings to the own demonstration of affection in the media during prime time.

At this level, there are several stigmas that exist, such as that homosexuals are futile and lecherous, and that they do not need to marry and have children. When discussing any issue about sexually transmitted diseases, homosexuals are automatically called to the subject. In contrast, little is said about them when discussing human rights (Irigaray & Freitas, 2013). The stigma that homosexuals and bisexuals suffer is due to their social identity, an identity that is invisible—their sexual orientation. Many adjectives are used to characterize male homosexuals, with a pejorative nature, which portray weaknesses, fear and inconsequence, thus justifying their fear of being stigmatized, because we are not aware of the impressions we make on others and the social reality they perceive may not be the true one (Irigaray, 2007, referred by Frazão & Rosário, 2008).

2.3 *Organizational Diversity Management*

Currently, the work environment in organizations is increasingly characterized by its heterogeneity. Different individuals live together for a long time in a single space where they carry out their professional tasks. In these environments, sometimes conflicts are created, and sometimes alliances are created, both between colleagues and with superiors.

Siqueira and Zauli–Fellows (2006) and Irigaray (2007) report that, with regard to sexual minorities in organizations, there is still a visible prejudice that can directly interfere with performance, job satisfaction, the level of stress and commitment of the individual with the organization.

With the growing diversity of people in organizations, administrations and/or top management needed to develop management strategies to bring profit, organizational practices and social justice into line (Robins and Coulter 1998, referred by Frazão & Rosário, 2008).

Both for employers and for employees, the coexistence with individuals of different social identities is a great challenge. This theme has been gaining importance in the scope of several studies and in organizational contexts.

About the management of diversity in organizations, Myers (2003) describes two major fields of studies: one with a focus on social responsibility and the other with a focus on competitive advantage. When competitive advantage is the focus, a greater understanding and need to know about diversity management are observed. However, and regardless of this, at the practice, there is little agreement between the proposed diversity policies and the policies that are really implemented.

Zauli-Fellows (2006, referred by Frazão & Rosário, 2008) highlights that diversity management in organizations requires specific skills, appealing to awareness, understanding and action. To implement a diversity policy, a real understanding of the diversity concept is needed, thus allowing the organization to improve both the team and the organization's performance.

That said, it is known that heterogeneity in the work environment challenges not only individuals but also organizations themselves.

The greater the diversity of human resources, the greater their creativity. It is easier to exchange information about experiences, values, attitudes and learning about new approaches, contributing to a better decision-making process for organizations. However, this organizational heterogeneity can also have controversial results such as the reduction of integration and social contacts, conflicts and communication problems (Irigaray, 2006, referred by Frazão & Rosário, 2008).

Siqueira and Zauli-Fellows (2006) state that when it comes to diversity policies, people are more likely to accept ethnic, social and gender differences; but they are resistant to the diversity of sexual orientations. According to these authors, there is a tendency for homosexual individuals not to reveal their sexual identity in the workplace, due to social pressures and harmful experiences previously had in the work context. Still, others adopt the strategy of hiding their sexual orientation and even faking a heterosexual relationship.

3 Methodology

The present study is descriptive-qualitative. The research is characterized as being exploratory since there is not enough data and literature that can serve as a basis for a research with regard to the Portuguese context. Systematized knowledge about homosexual identity in the workplace at the national level is scarce.

To better explore the proposed theme, in addition to the literature review, a semi-structured interview was carried out with two male homosexuals aged between 22 and 34 years old.

3.1 General Objectives

As general objectives, the present study looks to:

- Understand to what extent the discrimination of an employee, due to his/her sexual orientation, can cause emotional harm to him/her in the work context;
- Understand the extent to which discrimination against an employee, due to his/her sexual orientation, can cause dissatisfaction in the performance of his/her duties and even in the occupation of his/her job.

3.2 Specific Objectives

Established the general goals, and in order to achieve them, it is important to be more concrete, reason why we have defined the specific goals, namely:

- Identify the employee's perception of sexual discrimination by the organization where he works or has worked;
- Identify the employee's perception of sexual discrimination by peers; Explain the employee's disclosure and/or omission to the organization and peers about their sexual orientation.

3.3 Method

To develop this study, and in accordance with the established objectives, an interview script was elaborated and applied to two homosexual employees, who have been working in an organization for more than 2 years. Employees were asked to answer the questions based on their working life experience, not having to specify themselves in the company they currently work for.

The interview script was based on open questions about the perception that the interviewees have regarding the condition of their sexual orientation in their lives, at both a personal and organizational levels.

The interview is divided into 3 sections of analysis:

1. Personal and family identity;
2. Homosexual identity in general;
3. Homosexual identity at work.

This division is relevant because, as mentioned in the literature review, in order to understand diversity as such, it is necessary to know how the individual perceives diversity in general, as well as in individual terms.

4 Interviews' Presentation

Two interviews were developed with two homosexual individuals, aged between 22 and 34 years old. Both of them are male. The first respondent (E1) comes from a more rural environment, while the second respondent (E2) comes from a more urban environment, albeit from a small town.

With these interviews, we look to obtain feedback from the interviewees regarding their perception of the role of their sexual orientation in their personal and professional/organizational life.

Upon requesting these interviews, both respondents volunteered to be part of this exploratory study. They signed the informed consent and stated that they expected their testimonies to help in disseminating information regarding acceptance of the homosexual identity.

The interviews took place in a calm environment, at different times, lasting approximately one hour each. The recording of the interviews was authorized.

Regarding the interviewees, they are two young homosexuals who are assumed to the closest people.

The grids below summarize what was answered in E1 and E2 interviews (Table 2).

5 Interviews' analysis

The specific objectives were followed as a basis to analyze the interviews:

- ***Identify the employee's perception of sexual discrimination by the organization where he/she works or has worked.***

  E1, when referring to sexual discrimination directed by the organization, is perennial in stating that neither in the current organization, nor in the previous one, felt discrimination for being homosexual.

Table 2 E1 and E2 interview analysis

E1 - 34 years old. Professional in the telecommunications field. "It's a new area for me, I've always been a shopkeeper but always connected to fashion". Coming from a rural location, well-off family		
Personal and family identity		
Perception of his sexual orientation (how? When?)	**His everyday homosexuality and coming out**	**Family knowledge/perception and reaction**
"I realized it late, at 22, 23 years old, although I think it was born with us". Since I was born in a very rural and closed place and people intend to get married and have children, it seems that we are programmed for that, it ends up passing by you the discovery of our sexual orientation or ourselves". "I went out with heterosexual people with the definite idea that it was to get married and have children"; "I even had a relationship with a woman, the relationship ended due to circumstances of a normal hetero relationship, jealousy, lack of trust…"; On a summer night (…), with hetero friends, we're having fun on the track and I feel I'm being harassed, at the time I felt very uncomfortable but at the same time it hit and I didn't know if the nuisance was a bad thing being disgusted by the situation or because I was enjoying being harassed by a man. I ran away from that" "Later on a night of fun (…) in a gay environment (…) I didn't even know this bar was a gay space (…) glasses, we were all fun (…) When I enter the bar I realize that my gaze was only on the floor, I felt so observed that I felt uncomfortable once again. Realized that (…) once again being harassed. And once again I felt uncomfortable and left. That's when I started to realize that anything was not normal and that I had to have time to organize myself."	*"Things (the way of looking at homosexuality) have changed very quickly over the past few years and there is still a lot to change"; I am an open homosexual, I have no relationship at the moment. It's like that, I'm assumed to myself and my family, I'm not assumed like that "anyone knows", no, to me and to the people around me and who have meaning for me, which makes sense."; "Many people have this need for conversation to reveal (…) speaking in a sense with the family does make things easier"; "Because they think: it looks bad, there are people who are like that, only many years later (…) they will realize that it was a mistake not to accept it, that it is not worth it, that it is not a disease, nor that it was an option and that the person is really like they are like we are too."*	*"(…) until when she (sister) said "we have to talk" after about 3 weeks, I was in my room, she enters, closes the door, sits on the bed and says "look, you can start talking". Honestly, at the time, I just wanted to get under the bed because I didn't know what was going to come out of there, she was the first person who confronted me, I wasn't the one who had the initiative, she was the one who confronted me, however, I denied everything, I said: "I have nothing to say": "(…) and out of nowhere she says to me: "You're in love with a man", just like that in cold weather. There I started to cry, and she started to cry and started to talk, saying that, regardless of everything, I was her brother as always and that she was there for whatever came and went."; "And of course later we were talking about this subject saying that it was complicated that if you could choose it wasn't like that, that it was not an option that I could choose, that I always ran away from feeling it, but that it was no longer possible to hide it." "I can tell you that I never had a conversation with my parents like that. I've already pondered, already, I've talked to my sister about this to which she told me "You feel this need to talk to our parents" and I "no, I don't feel this need". I don't know if I'll have it in the future, not right now."*

(continued)

Table 2 (continued)

"(…) The first relationship I had was when I had a store and was a customer. In a very polite and calm way he knew how to reach me and after a month of insisting I agreed to go to dinner with him. And it started then, there I still hadn't accepted me, I was still accepting me. I only accepted when my sister, who is the closest person to me and who knows me without me talking, just looking at me, one day she said: "one day we have to talk". At the time she had never talked about it but when she said that to me I already knew that she wanted to tell me something about it. At that time I should have around 27/28 years old"	*"I am aware that I am a very open and well resolved person, and that not everyone who is homosexual thinks that way (…) I also have some fears, I am not 100% comfortable with myself, it doesn't mean that I accept myself, I mean accepting I already accepted myself a long time ago (…) but like at ease, the relaxation of being who I am I still have these fears (…) I still have this fear, but I live well with myself right now."* *"I wasn't capable of holding hands in public, of giving a kiss. I know myself and I am not capable of it. I also think that because I'm a reserved and shy person, even in a hetero relationship, I've never done that, but I also think I wasn't comfortable with being gay and society itself still isn't at the level a person would want. I think that this is why I couldn't (…) Being all equal, that we're not yet and we won't be anytime soon, at least not here in Portugal."*	*"I had two relationships to live with the person I was with. People (boyfriends) would go to my house for dinner and my parents would even say "look invite to lunch" or "invite to dinner", to come to the birthday."; "They accept my friendships, they knew I slept with the person in an apartment, so my parents are not naive, to the point of not knowing…they know…I'm their child, they know me and they have the perception that I eat things they are with me, that I am different or that I am the same"* *"As there wasn't such conversation, I can't be sure, however, I think so because there were situations where my ex were at home and so my parents always treated them the way they treat my brothers-in-law. Normally, my parents treat their sons-in-law for you, and they also treat the people I've related to as you. It's funny that…while my friends my parents take care of you*

(continued)

Table 2 (continued)

Homosexual identity in general	
Characterization of the homosexual population	**Homosexuality today by the general population**
"I don't deal much with the homosexual community, I'm not very gay, I've never been, I feel better with hetero. It's a funny thing, I give myself, I have gay friendships, but if it's in a gay environment I think that gays themselves lack respect for themselves, you know? And I think hetero when I'm set with hetero there's always a distance, a respect, and there's friendship in it, and gays always seem to have an ulterior motive, I don't know, it seems like after friendship there's always a way to overcome that and, how can I explain, something happens, some physical involvement, anything is a reason for involvement. They prefer sometimes to spoil a friendship for a moment. And I'm not like that, that's why I think sometimes I walk away because it's so easy in the gay world, it's so easy to get past this barrier, it bothers me. That's why I prefer the hetero part, a more hetero environment, for that reason, because I know it's not easy, I'm more relaxed, in a gay environment it seems I'm always on the defensive, anything is a reason to get closer. The gay community is like that, there is a certain promiscuity, maybe because it is a very hidden world, and these environments are their alone, a world still very closed, as it is hidden it is easier"; "Yes, still, here in Portugal still yes. Because if you go abroad, I was in Germany for a year and I didn't notice it. I walked down the street and saw homosexuals holding hands. And myself, if I see here in Portugal, a gay couple walking hand in hand, I look at myself with that look of admiration"	*"I think they are increasingly accepting the idea that being homosexual is a normal thing, but there is still a lot of homophobia in this regard… still…I feel…in fact, the simple discomfort of seeing someone, myself, on the street, holding hands, giving a kiss, gay, go around our society a little, I feel that society feels, I have a little the pain of society, because I also have that fear because I can also be on that side. Therefore, I feel that society is still far from what a homosexual would want."; Expressions like "oh gay; you're a sissy (…) I feel that this is a way of playing with the situation because if you talk to the person in question, sometimes that person has no prejudice, sometimes there are things from many years that those expressions remain over time that they do not transmit homophobia but are expressions that remain without malice."; "I think homosexuals don't show society what they really are. The homosexual is not a clown who wears strange clothes and who walks down the street kissing and making great demands, being ridiculed"; "(…) in public opinion, it seems that a homosexual is not a man, it seems that he is a clown who walks there… ridiculing in the worst way the image of a gay person. That's why hetero people, our parents, when they look at this spectacle, will portray their son, or grandson, or neighbor with that image and that's not true, it's not the one that corresponds. Homosexuals are normal people, just like the rest of society. They don't have to go around saying they're gay, I don't go around saying I'm gay, hetero people don't go around saying they're hetero either. I walk in the street, I wear normal clothes, like the man I am, my attitudes are that of a man, the gender that characterizes me, now I don't have to reveal that I'm only gay because heteropeople don't do it either."; "As I belong to the society where we are, I live, and filter everything…and now it's living day-to-day. Now, in the sense of having someone, I still have a long way to go to feel safe and at ease."*

(continued)

Table 2 (continued)

Homosexual identity at work			
Homosexuality in the current work context	**Do colleagues and bosses know? How did they know? Why this need?**	**Feeling of discrimination at work**	**Importance of diversity in organizations**
"In my opinion I think it's a part that has never really worried me that much, honestly. Whether you're gay or hetero, for the entrepreneurial part, I even think there are advantages, and I think some feel that advantage in the sense that a gay person normally doesn't have children or not want to build a family. You can even have one, but it's a rarer thing. That is, there won't be maternity leave there (…) the current prejudice, more than being gay, is a woman getting pregnant and being fired, for me the worst discrimination is this. And for me, the entrepreneurial part discriminates more in this type of situations than the sexual orientation part."; "I think the law defends the difference. Although I've never gone through that, I think that if I had to go through a situation of discrimination, the law would be in my favor, even because I wouldn't have any fear of going ahead with the law because I think I'd win this fight."	*"Yes, they do. At least those I think should know. Some I told, others asked me and that's basically it. Even to get to know each other, just as I asked, they also asked. It was for us to understand each other and the revelation of the person I am emerged. Now, if I have homosexual characteristics? I must have, sure, because I think that every gay man, however masculine he is, there is always a point there or the sensitivity or the way of looking, or the way of dressing, or the way of reacting… there is something there, or thinking even, something there stands out and it is noticed sooner or later who lives with that person. Of course, there are people who, just by looking, if there is no coexistence, it is not discovered that they are gay*	*"In my professional experiences, I never even felt the need to talk about this personal part of me. I've had bosses who knew about me, and who are now great friends of mine. The treatment was always the same and I think it was even better when they knew my sexual orientation because when we give ourselves, we give 100% to the person, the person realizes that we are being us and genuine." "(…) when the person realizes that we haven't revealed that part, they know that we are not being ourselves, isn't it? Something is missing… especially the main one, if you don't say who you are, the person will reach a point of conversation… there will be a barrier there. If we are genuine, we are no doubt accepted for our differences, no doubt."*	*"I think it's very important to have diversity in organizations. We are talking about people who are different, but at the level of different personalities, this is all part of it. In fact it is as important as a man working with a woman, instead of working only men or only women, it is the same thing."; "(…) they are different personalities, different sensibilities, the organization gains from this… you always learn something, because the way of thinking will also influence the other person"; "(…) gay people are more sensitive, more careful, more detailed in the sense of whatever task they are doing (…) Hetero people are perhaps more practical, more concrete in tasks… perhaps they will also add something to their colleague gay. This is just an example, which sometimes has a little to do with the personality, the main difference in these two men is really what I said."; "Usually this is what happens and maybe it will influence and add something to the other"*

(continued)

Table 2 (continued)

		"I was never discriminated against, neither by colleagues nor by supervisors. I don't remember having any situation in which I felt uncomfortable, or even discrimination. I never had this experience, fortunately, I don't even remember people who had discriminatory situations."; "However, I know that discrimination exists. In the organizations where I worked and as I was a manager and, by chance, I didn't manage anyone like me, but if I need to, it will be the same to anyone who was working there, because usually I don't mix work with personal life. That is, I can... there are more relaxed days, more tense days, more professional days, that we have to be there for work. But with both things I can't mix. When I was a manager, I was always professional and I never gave my personal side to situations. So I think that if it was someone else like me, I don't think there was any impediment. In fact, today I'm a colleague of a person like me and there's respect, we happen to get along well outside of work, there's respect in it. There it is, I don't think there's this discrimination, we know our place, and it's keeping it with respect	

(continued)

Table 2 (continued)

"Honestly, with regard to this issue of discrimination, I can't understand. That's why I'm not being able to answer your question very well, because I don't understand well the boss and the prejudice of homosexuality. Maybe because I never lived that, but if you want to know, I didn't even admit it. I didn't admit that lack of respect, nor did I allow for it. Before we were different, we are people and we should be respected. We should respect and we should give ourselves to the respect. If you want to know, I don't even have patience to discrimination. And I'm not just talking about gays, I'm talking about other ethnicities, gender differences. It's no longer a matter. These people who don't accept it, nowadays, they no longer have room for not accept. They may not accept unconsciously. But now… at the level of public… to show… and so… I don't even have the patience for that anymore. That's why I think if it happened to me or if I saw it happen, I wouldn't admit it."			*"I think it's important, the demand for gay rights…, it shouldn't be important (…) it should be normal because we are all the same, but as this is not the case, I think it's also important for people to change their mentality"; "We are in need of shaking up and dealing with these issues as a non-issue. To show that, for example, homosexual relationships are normal relationships. Oftentimes, there is more abnormality in heterosexual relationships, due to the disagreements and discussions that arise out there."*

(continued)

Table 2 (continued)

E2 – 23 years old. Work at a telecommunication store, serving the public. Coming from an urban area		
Personal and family identity		
Perception of his sexual orientation (how? When?)	**His everyday homosexuality and coming out**	**Family knowledge/perception and reaction**
"Since young, since I started to have more access to the internet, mainly, I started looking for and enjoying seeing men more than women and the internet was really a driver and a facilitator" *"At the time I liked to see it physical attraction, in fact you were more aware that it was physical attraction than sentimental, that's what I thought"* *"Age 12/13, I already had these ideas, but it was more serious from 16 years onwards. More seriously means realizing that I was more attracted to men. Not in feeling, because having feelings for a man happened later."* *"I have, I have, assumed in front of people who are close to me and who have meaning for me. It is a recent relationship, we are still in the beginning, to see what happens, but the people around me are aware of this relationship"*	*"I'm an open homosexual but who knows are the closest people, I'm a very open person, I get along easily with people. I don't have problems saying that I'm a homosexual nowadays. There are other people I prefer not to know because I don't get along with them, they are not my friends so I prefer they don't know. (…) it's not hiding because it always ends up being known, but it's a matter that concerns me only. A hetero person doesn't go around saying he's hetero, does he?"*	*"My mother is French, average, came to Portugal alone and was educated by my great-grandparents who have passed away. My father was always from a poor family, there were many children. Then he got together with my mother, it's me and two other sisters"* *"From a very early age my mother started to realize this, through the internet history and some day she confronted me with these things. At the time it was not good, isn't it, I don't know if it was because it was by men, or just because it was pornography. I didn't quite understand, but at the time my mother didn't handle it very well. My father is a very closed person, I don't know, I never talked to him much about this. It's not hidden or revealed, it's just something we don't talk about. I see that my mother, as she is the manager of an entertainment establishment, and the diversity of customers is great, maybe she accepts better. Everything is more relaxed, but now I don't know how different I am either. I don't know if in relation to her son, things are so calmly accepted"*

(continued)

Table 2 (continued)

	Coming out? I know the expression very well and I never had the opportunity to say: "I'm a homosexual mother". I was discovered. But I see, for example, several videos on the Youtube of couples who film themselves "coming out of the closet" and revealing to their parents that they are gay, filming their reactions *"When I say that I'm a homosexual I always hope that people accept me naturally. I don't need to be accepted but it should be because I don't like to get along with anyone, nor do I want a person to look at me sideways or leave me alone speak for being homosexual"* *Yes, I've done it, not in my homeland but in places where I know the mentality is more open because I know there are already a greater number of gay people on the street and yes I have no problem walking hand in hand or kissing in public"*	

(continued)

Table 2 (continued)

Homosexual identity in general	
Characterization of the homosexual population	**Homosexuality today by the general population**
"Homosexuals have a more animated euphoria characteristic, they are more outspoken, friends and I think that's good to have in the context of the work environment." *"More and more people are accepting homosexuality. I just think that there is this group of homosexuals, who like to show off, who like the extreme of tics and show offs, which makes people talk bad and not like homosexuals"* *"There are many split groups. There are homosexuals who I think are normal people, a man who likes a man, to be perfectly masculine. And I think there is that group of homosexuals who like men but who also like to show off, who like to be the center of attention, that are those called queers and that I personally do not like and do not agree with their postures. But that's it… queers…"*	*"People are educated in different ways. Some more open education, others more closed. Homosexuality is seen as something different. It has a lot to do with education. In all cases where there is difficulty in accepting a homosexual, it has to do with the education that these people had. If a person from an early age is educated to say: it's not normal, you can be what you want to be, you can like whoever you want, regardless of whether you're a man or a woman, you don't have to choose to be hetero just because men have to be with women"* *Homophobia - "I think people who do that do it when they're in groups of friends. They always try to feel superior to the group of friends, because I think if we're going to confront these people directly when they're alone, I'm talking in people with my age, I think if we present them all the facts, what's wrong, what's wrong with homosexuality, I think people aren't going to be what they are when they're with friends."* *"For me homosexuality is considered a normality. But I understand other people when they say that it is not normal for a man to like another, I speak of normality as being usual. It is unusual to see a couple of two men, and normally, one thing that is different from others in the minority people tend to say that it is not normal, that it is different, although we are all people."*

(continued)

Table 2 (continued)

homosexual identity at work			
Homosexuality in the current work context	**Do colleagues and bosses know? How did they know? Why this need?**	**Feeling of discrimination at work**	**Importance of diversity in organizations**
"I think people don't look at that, in top management. I think homosexual people don't hide that they are. In my company I have homosexual bosses, I know that one of them is open, the other is not. Therefore, I think that the organization where I work, they do not see the fact of being homosexual as an impediment to progress in their career or to work or not in the organization" *"Homosexuals have a more animated euphoria characteristic, they are more outspoken friends and I think that's good to have in the context of the work environment."* *"I prefer to be honest and at ease and I know that with my colleagues I can feel that"*	*"Perfectly normal, most colleagues know that I'm homosexual. My boss knows, but, for example, there is a colleague who doesn't know, but only because I don't deal with him much. Ok… he's my colleague, is part of my life but he's not my friend and therefore I don't have a conversation and that's why (…) we don't talk about it. However, if he comes to me one day and asks me, I'll tell him yes I won't lie"*	*"No, not for a while, I've never had this confrontation with discrimination, or at least I didn't pay attention to it, being irrelevant to me. If we talk about school days, it was different, but not in a work context. People are adults and have to coexist with each other. But like this, discrimination that is not known"* *"I think that people have to take the credit they have for the work they do, not for being homosexual or not, because each person's private life or their sexual orientation only concerns the person."*	*"I think that a manager doesn't have to think: "I'm going to put him in because he's gay, or I'm not going to put him in because he's gay", no, I'm going to put him in this position because he works (…) because he's good at what he does"* *"I hope it doesn't happen in the mind of a manager, not putting A or B in a position or even not hiring because I'm gay. I don't know, I don't know about it. I really hope not, but if a person is a worker and is suitable for the job, I don't think that the fact of being a homosexual interferes with that"* *"I think that at a business level, mentalities are changing. It has to do with changing the mentality of people in general. And… that's it… when they are in the environment where people get along, there is no reason to discriminate against each other"*

(continued)

Table 2 (continued)

	"First, because I like to talk freely with people, and the fact that colleagues ask: do you have a girlfriend or not, or talk about women who pass in front of the store, it would be a bit of a lie, (…) to align with my heterot colleagues *"My boss (…) is perfectly equal to the treatment he has with my colleagues. Also because I had the possibility to move up in the service line and it was proposed to me by the boss"*		

> *"I was never discriminated against, neither by colleagues nor by supervisors. I don't remember having any situation in which I felt uncomfortable, or even discrimination. I never had this experience, fortunately. I don't even remember people who had discriminatory situations."*

> *"However, I know that discrimination exists. In the companies where I worked and as a manager and, incidentally, it wasn't from anyone like me, but if I were, I was the same as anyone who was working there, because normally I don't mix work with life folks..."*

- E2, referring to sex discrimination by the organization where he works, reveals that he does not feel any type of discrimination.

> *"No, not for the time being, I've never had this confrontation with discrimination, or at least I didn't care about it, being irrelevant to me. If we talk about school days, it was different, but not in a work context. People are adults and have to coexist with each other. But discrimination that I am aware of, no."*

- ***Identify the employee's perception of sexual discrimination by peers***

  Regarding the perception of discrimination by peers, E1 referred that he never had the perception that his colleagues had some kind of problem for being homosexual.

> *"I was never discriminated by my colleagues. As I told above, I don't remember having any situation in which I felt uncomfortable, or even discriminated. I know that discrimination exists. But, fortunately, I never had this experience".*

In what concerns E2, he reveals that so far, he still hasn't noticed any type of discrimination by his colleagues. He says that as adults they are, it makes sense to live with each other's differences.

Asked whether there is physical visibility for homosexuality, respondents replied that:

> E1—*"Now, do I have homosexual characteristics? I must have for sure. Because I think that every gay person, no matter how masculine, there is always a point there, or the sensitivity or the way of looking, or the way of dressing, or the way of reacting. There's something there, or something to think about. Anything there stands out and it becomes apparent, sooner or later, who is related to that person. Of course, there are people who, just by looking, if there is no coexistence, it is not discovered that they are gay."*

> E2—*"Homosexuals have a characteristic of euphoria, more animated. Are more foul, friends...".*

Asked about homophobia:

E1 referred that expressions such as,

> *"oh gay; you're a sissy (...) I feel that this is a way of playing with the situation because if you talk to the person in question, sometimes that person has no prejudice. Sometimes there are things from many years that those expressions remain over time, that they do not transmit homophobia, but are expressions that remain without malice".; I think homosexuals don't show society what they really are."*

E2 explained that,

> *"I think that people who do that do it when they are in groups of friends, they always try to feel superior to the group of friends. Because I think if we are going to confront these*

people directly when they are alone, I am talking about people with my age, I think if we present them all the facts, what's wrong, what's wrong with homosexuality? I think people won't be what they are when they're with friends. I think it's really a matter of straight men in the group trying to..."

– ***Explain the employee's disclosure and/or omission to the organization and peers about their sexual orientation***

E1 revealed to be homosexual to the one who asked him, and others who knew because he thought they should know. He said it happened naturally, on occasions that allowed him to get to know his colleagues better and vice versa.

> *"Yes, they Know. At least those I think should know. Some I told, others asked me and that's basically it. Even to get to know each other. Just like I asked, they also asked. It was to understand each other and came the revelation of the person I am"*(E1).

E2, regarding the reason why he told about his sexual orientation, at work, said that:

> *"First, because I like to talk freely with people, and the fact that colleagues ask: do you have a girlfriend or not, or talk about women who pass in front of the store, it would be a bit of a lie, (...) to align with my heterosexual colleagues. I prefer to be honest and at ease and I know that with my colleagues I can feel that."*

However, both interviewees also mention that they never felt the need to say that they were homosexual when they presented themselves in a new job. Respondents indicate that this matter only concerns them. It is individual, and when they go to work it is to be professional, and not take his personal side to work.

> E1— *"In my professional experiences I never felt the need to even talk about this personal part of me. I've had bosses who knew about me, and who are now great friends of mine. The treatment was always the same and I think it was even better when they knew about my sexual orientation because when we give ourselves, we give 100% to the person, the person realizes that we are being us and genuine." "(...) when the person realizes that we have not revealed that part, they know that we are not being ourselves".*
>
> *"(...) If we are genuine, we are no doubt accepted for our differences, no doubt."*
>
> E2— *"I'm an open homosexual but who knows are the closest people. I'm a very open person. I get along easily with people. I don't have problems saying that I'm a homosexual nowadays. There are other people I prefer not to know because I don't get along with them. They're not my friends, so I'd rather they don't know. A heterosexual doesn't go around saying he's heterosexual, does he?'*

Both interviewees state that for employees to have a good performance it will not be because of their differences, being gay or heterosexual. It will be because of their effort, work and dedication. Of course, the individual who suffers discrimination within the organization is emotionally fragile, which can compromise the good performance of their functions in the organization where they work. They also refer to the added value of diversity with regard to the environment created that facilitates quality of life at work. They referred that, for example, homosexuals, because of their more extroverted and/or more sensitive personalities, balance the environment among teammates, even providing managers with the use of skills and talents of workers. Both end with the thought that organizations should not suffocate the worker and their ideas because of the existing diversity, but take advantage of what different people can contribute to the organization.

This analysis of the interviews suggested some facts different from that of the literature review; namely, the interviewees do not perceive discrimination and indicate that this difference is felt less and less in the work context. They explain that, since sexual orientation is something individual and intimate, it does not have to be revealed in a work context, as this characteristic will not imply better or worse performance of their functions. Regarding homophobia, they always highlight the opinion of people who have no meaning for them, as such, it is not important for them, whether individual A or B comment on their orientations. In short, it appears that there is a long way to go in academic studies and researches, and not only, with the objective of dealing with diversity not in its theoretical form, because at this level there is great consolidation, but at the level of practices which are not written or revised, in order to facilitate the management of diversity by our organizations, taking into account the Portuguese society.

6 Final Remarks

It is hoped that this chapter, and as it has always been referred to throughout it, encourages further research at the academic level with the aim of deepening the theme of diversity in Portuguese organizations. Encourage studies with greater objectivity in order to suggest the direct and indirect results of greater diversity in the context of organizations. For instance, understanding which means can be used to achieve greater individual performance using good diversity management practices, and understanding which are the bad practices to learn from mistakes, resolve them and not remedy them. It is suggested that a survey about the diversity management policies practiced in our organizations should be carried out, that cover not only issues of race and ethnicity, but also other different forms of inclusion, such as, in the case of this chapter, sexual orientation. Emphasize sexual diversity and understand how it can lead to positive or negative changes in the individual's performance, in a work context, related to the management that is done and the attention received.

Another suggestion has to do with the mobilization of research in the scope of diversity policies that belong to the public sphere, transporting them to the sphere of organizations, in a process of social privatization. It is very important that people get rid of the prejudiced looks that allow bias, and the looks that block the importance of difference, not only in social life but also in work. That organizations see the difference as part of talent and human potential.

Last but not least, we must re-emphasize diversity and the concern of its management by organizations. Continue to raise awareness with top management, using the studies and research that are being carried out in this area. It may seem like a natural and reductionist subject, but, it is a fact that these issues are "normalized" not to be talked about. The issue of diversity may often seem well discussed in society in general and in organizations in particular, but that is not what is currently happening.

In this chapter, it was verified through the analysis of the interviews of both interviewees that despite knowing that discrimination exists, both in the work context, and in general, they admit that they never felt any kind of discrimination. They agree that it makes sense to talk about diversity and its management by organizations because, in a work context, the opportunity must be given to the individual, to their individuality and to the ability to socialize.

The differences of each one should be a triumph for organizations to capture talent and encourage their employees to progress in their careers. This is the best way to manage diversity in organizations. However, there is still a lot of resistance for managers to think and act this way.

It is also concluded that it no longer makes sense to talk about discrimination when, nowadays, it should no longer exist—at least when we mention it in its pejorative form. Respondents said that it exists, that sometimes it is even hidden, but that mentalities are changing and that is why we must elucidate people, both employees and managers.

The mistake is to think that everything has already been done and thought about and that, if the management of an organization results in a certain way, there is no reason to change.

Diversity is only considered when socially inserted, however, interpersonal relationships depend on how individuals themselves perceive their integration in the social environment.

References

Costa, A. B., & Nardi, H. C. (2015). Homofobia e preconceito contra diversidade sexual: Debate conceitual. *Temas Em Psicologia,* (3), 715–772.

Dinis, N. F. (2011). Homofobia e educação: quando a omissão também é signo de violência. *Educar em Revista,* January/April(39), 39–50.

Frazão, P., & Rosário, R. (2008). O Coming Out de Gays e Lésbicas e as Relações Familiares. *Análise Psicológica,* 1(XXVI), 25–45.

Irigaray, H. A. (2007). *Políticas de Diversidade: Uma Questão de Discurso? Anais 25º ENANPAD.* Brasil.

Irigaray, H. A. (2006). Prejudice against lesbians, gays, and bisexual employees and its impacts on their health and well-being. *Anais 26th International Congress of Applied Psychology.* Atenas, Grécia.

Irigaray, H. A., & Freitas, M. E. (2013). Estratégia de Sobrevivência dos Gays no Ambiente de Trabalho. *Revista Psicologia Política,* (26), 75–92.

Myers, A. (2003). O valor da diversidade racial nas empresas. *Estudos Afro-Asiáticos,* (3), 483–515.

Pachankis, J. P.& Goldfried, M. R. (2004). Clinical issues in working with lesbian, gay, and bisexual clients. *Psychotherapy: Theory, Research, Practice and Training, 41*(3), 227–246.

Rohm, R., Fernandes, F. A., & Dantas, C. A. (2012). Diversidade Sexual e Vantagem Competitiva: Uma análise de Seus Impactos nas Organizações. *Anais Do XXII Encontro Brasileiro De Administração e VIII Congresso Mundial De Administração, 01,* 373–388.

Siqueira, M. V. S. Zauli-Fellows, A. (2006). Diversidade e identidade gay nas organizações. *Gestão.Org., Revista Eletrônica de Gestão Organizacional4*(3), 69–81.

Wilson, I. (1999). The emerging gay adolescent. *Clinical Child Psychology and Psychiatry,* (4), 551–565.

Zera, D. (1992). Coming of age in a heterosexist world: The development of gay and lesbian adolescents. *Adolescence,* (108), 849–854.

Corporate Social Responsibility: From the Origin to the COVID-19 Pandemic

Carla Magalhães

Abstract Corporate Social Responsibility is a concept that dates to the first half of the twentieth century, although only in the twenty-first century it has become a paradigm for most companies. The concept itself is imbued with a certain versatility, as there are countless authors who theorize about it, as well as several theories that attest to its nature, with the work of Garriga and Melé (J Bus Ethics 53:51–71, 2004) being one of the most expressive in that sense. However, to understand the importance of Corporate Social Responsibility and why the concept has been disseminated throughout history, it is pertinent to refer to the advantages that it brings when implemented by organizations, that is, it is pertinent to know the motivations that lead companies to invest in Corporate Social Responsibility. This issue becomes even more relevant in a scenario of crisis, such as the one triggered by the COVID-19 pandemic. Given this, it is urgent to ask: how are companies assuming their social responsibilities in this pandemic scenario? What guidelines are they following? Are companies aligned with the current needs of their stakeholders?

Keywords Corporate Social Responsibility · COVID-19 pandemic · Organizations

1 Introduction

Organizations are currently facing major challenges, which have promoted new trends, as a result of the current scenario, marked by great unpredictability and volatility in terms of organizational management, with particular emphasis on industry. One of the biggest challenges and trends for organizations is related to social responsibility practices, which should be seen as a model to be followed, in favor of the balance and well-being of all parties involved in the organizational activity. Indeed, these practices can and should be viewed strategically, as they promote the company's competitiveness while benefiting the several actors involved with it. That

C. Magalhães (✉)
The Transdisciplinary Research Center of Innovation & Entrepreneurship Ecosystems (TRIE), Universidade Lusófona Do Porto, Porto, Portugal
e-mail: carla.magalhaes@ulp.pt

C. Machado (ed.), *Challenges and Trends in Organizational Management and Industry*, Management and Industrial Engineering,
https://doi.org/10.1007/978-3-030-98048-1_4

is, for an organization to be competitive, it is essential that it operates in an environment that promotes balance between its interests and the interests of its stakeholders, which becomes even more important in a critical scenario, marked by a pandemic that has very negative implications for organizations that are unable to adapt to new market rules and practices. And if previously socially responsible practices could be seen and applied in an isolated and sporadic way, currently the recommendation is that they act in an integrated, continuous, and sustainable manner. Only in this way companies can build solid production structures, based on contexts of trust and mutual help, contexts that are so important in times of crisis.

Therefore, it is important to analyze the concept of Corporate Social Responsibility, going back to its origins and deepening the advantages that its application can bring to organizations. It is also interesting to carry out this analysis in a context marked by COVID-19, to understand whether this pandemic has been an obstacle or a driving force for socially responsible practices by organizations. Understanding the contours of this issue within the current scenario is of paramount importance, so that organizations can make strategically sustainable decisions and contribute to an effective response to the crisis created by the pandemic, which cannot be overcome without concerted and conscious action of the business world. In short, organizations must be aware of their impact on society and on environment, their capabilities to make it more positive and the possibility of this impact be productive for their own business.

2 The Concept of Corporate Social Responsibility

The concept of Corporate Social Responsibility (CSR) is very broad and may acquire different dimensions depending on the context in which it emerges, namely the level of development of the country where it is applied (González-Masip, 2021; Windsor, 2019).

To describe the different perspectives that involve the concept of CSR, we turn to the work of Garriga and Melé (2004) who refer that theories relating to socially responsible practices can be divided into four large groups: instrumental, political, integrative, and ethical. According to instrumental theories, companies are an instrument to create wealth; and therefore, CSR should be used solely for this purpose. The authors who stand out the most in this line of thinking are Levitt and Friedman. According to Levitt (1958), efforts to increase CSR, without having as their main objective their financial development, lead to the destruction of the capitalist system. In turn, Friedman (1962) argues that a company's social responsibility should only be based on the profitable use of resources and involvement in activities leading to increased profits if it follows the "rules of the game." Other authors who support this current are: Drucker (1984), who defends the idea that profitability and responsibility are complementary notions, as well as the idea that it is desirable for businesses to "convert" social responsibilities in business opportunities; Stroup and Neubert (1987), who also see CSR as an investment, which, according to the authors, can ease shareholder resistance; Jones (1996), who states that socially responsible behaviors, devoid of an instrumental perspective, are not compatible with the rationality of a capitalist company, fundamentally concerned with capital accumulation; Knautz (1997), who argues that companies should only be concerned about being productive and generating profits, as that is the only way they will be responsible to society; McWilliams and Siegel (2001), who advocate that decisions involving CSR should receive the same treatment as any other investment decision; Porter and Kramer (2002), who argue that CSR only makes sense if it ensures a competitive advantage to the company; and Keinert (2008) who states that managers have a single obligation, that of maximizing profits for shareholders, as if they had an employment contract as agents of the company's owners, among others.

As for political theories, the company interacts with society and assumes responsibilities in the political arena, using CSR as a vehicle for this interaction. In this perspective, we can highlight Bowen, for whom CSR is defined as businessmen's obligations to pursue policies, make decisions, or follow the lines of action that are desirable in terms of objectives and values of our society (Bowen, 1953). Mcguire (1963) reinforces the ideas defended by Bowen, as he states that companies must move toward accepting the social obligations and responsibilities arising from their prominent and powerful position in society. Consequently, certain social obligations or responsibilities arise from this finding, being also applied to groups and institutions and empowering companies that act in this way (Davis & Blomstrom, 1975). Altman (1998) refers to the concept of corporate citizenship, as the basis of corporate community relations and defines it as the function through which the company intentionally interacts with non-profit organizations, groups of citizens, and other stakeholders at the community level. Indeed, there are several authors who refer to the concept of corporate citizenship (Hemphill, 2004; Maignan & Ferrell, 2000, 2001a, 2001b; Matten et al., 2003; Peterson, 2004; Warhurst, 2001; Wood & Logsdon, 2002), a concept that integrates political theories, as companies have a certain power and influence in society and must manage this situation in a "citizen" way.

Integrative theories imply that the existence and growth of companies depend on society, so that CSR is a way for companies to integrate society's demands into their management processes, to legitimize their continuity. Here, the work of Carroll and Schwartz stands out, who present a circular model of CSR that includes three components (economic, legal, and ethical), which intersect with each other, with no one predominating over the others, but one overlap (Carroll & Schwartz, 2003). Thus, it is expected that the company produces goods and services and sells them with profit (economic component), respecting the law (legal component) and behaving ethically and respecting the norms (ethical component). In this regard, a call for attention to the fact that there are studies that indicate that, at times, social responsibility can be confused with its legal component, since there are companies that assume themselves as socially responsible when, in fact, they are just limited to complying with their legal obligations, which does not make them companies that promote CSR, as this must go beyond the law (Magalhães, 2011). Prakash Sethi proposes a classification into three levels, according to the company's behavior toward social needs: social obligation, which is summed up in meeting legal needs; social responsibility, which guides the company's action toward a congruence with social values; and social responsiveness, which presupposes a capacity to anticipate long-term social issues (Sethi, 1975). Another author who is concerned with the issue of CSR from an integrative perspective (albeit in a gradual way) is Frederick (1986, 1994, 1998), who presents four levels of CSR. At level 1 is the idea that CSR imposes on companies an obligation to work to improve social issues (Frederick, 1986, 1994). At level 2, Frederick (1994) presents the idea of corporate social "responsiveness," which can be defined as the capacity of a company to respond to social pressures. At level 3, Frederick (1986) incorporates the notion of moral correctness in the actions and policies formulated by an organization. At level 4, Frederick (1998) encompasses Cosmos (C), Science (S), and Religion (R). Until this is achieved, the company must be socially responsible (CSR1), respond to social needs (CSR2), and act with ethical and moral integrity (CSR3), integrating society's expectations into its management practices, as only in this way they are legitimized. This integrative current is also based on the so-called Theory of Stakeholders, which argues that all parties interested in the company's activity (such as employees, customers, suppliers, the Government, and the community), and not just its shareholders, should be considered (Evan & Freeman, 1993). This idea is reinforced by other authors (Bendheim et al., 1998; Berman et al., 1999; Donaldson & Preston, 1995; Freeman, 1984; Harrison & Freeman, 1999; Hummels, 1998; Logsdon & Yuthas, 1997; Mitchell et al., 1997; Phillips et al., 2003; Stoney & Winstanley, 2001; Varadarajan & Menon, 1988; Vos, 2003). Companies increasingly believe that CSR does not only involve respect for the environment, but they must also be economically and socially responsible, not only considering profitability, but also other aspects related to their stakeholders (customers, suppliers, collaborators, public authorities, and the general public) (Hutchins & Sutherland, 2008).

Finally, in relation to ethical theories, the company must relate to society based on ethical values, so CSR is seen as an ethical obligation. Within the scope of this theory, we can refer to Mulligan (1986) and Wood (1991), for whom companies should practice good deeds and not just avoid or prevent damage. For the authors, only the moral sense should guide the conduct of organizations, since the law is not enough to do so, as there are good laws and bad laws. Following these ideas, Enderle and Tavis (1998) present a model in which corporate responsibilities in the economic, social, and environmental dimensions should be considered at three levels of ethical challenges that corporations propose: level 1—where there are minimum requirements ethical—level 2—in which obligations beyond the minimum ethical level are considered—and level 3—in which there are aspirations for ethical ideals. According to Kok et al. (2001), the company must be socially responsible to society as a whole and must do it in a disinterested way, that is, without wanting to gain anything in return.

Finally, Mintzberg et al. (2002) also take an ethical stance, as they claim that companies should focus on the ethical values of society and not just on the sustainable management of economic, environmental, and social factors.

But how and when did CSR arise?

2.1 The Origin of Corporate Social Responsibility

Howard Bowen (1953), American economist, is considered the forerunner of social responsibility, due to the publication of the book "Social Responsibilities of The Businessman" where he questioned the responsibilities that a businessman should assume toward society and the decisions taken by entrepreneurs of various businesses that affected the lives of citizens. But even before that, it is possible to identify the first major highlight, in 1946, when Fortune Magazine questioned business executives about their social responsibilities—93.5% of respondents agreed with the statement that the so-called businessmen would be responsible for the consequences of their actions in a broader sphere than that of the shareholders they represent (Carroll, 1999).

Hay et al. (1976) give us a contemporary view of the emergence of CSR, stating that it developed in the United States, in the first half of the twentieth century. According to the authors, the evolution of CSR in the United States went through three distinct phases: Phase I, which was the management phase for profit maximization. The prevailing belief was that company managers had only the objective of maximizing profit; Phase II was the phase of trusteeship and emerged in the 1920s and 1930s of the twentieth century, as a result of structural changes in business institutions and society. According to this concept, corporate managers were responsible not only for maximizing shareholder wealth, but also for maintaining an equitable balance between the demands of consumers, employees, suppliers, and the community; Phase III, the management for quality of life, based on the significant metamorphosis of social goals in the United States. At this stage, the socially responsible company then becomes deeply involved in solving society's greatest problems (Hay et al., 1976).

Fry et al. (1982), in turn, refer that it was in the 1960s that companies, pressured to demonstrate their CSR, began to create their own foundations, inaugurating the phase called corporate diplomacy. This moment was characterized by the pressures imposed by stakeholders on companies, which, to avoid further regulation by the government or to obtain concessions, were forced to take administrative decisions that did not necessarily contribute to their profit, but that somehow could help them in future (Varadarajan & Menon, 1988). Hood (1998) states that CSR was already accepted as a doctrine in the US and Europe in the nineteenth century, when the right to conduct business in a corporate manner was a matter of State or Monarchy prerogative and not a private economic interest. Monarchs authorized charters for publicly traded corporations that promised public benefits, such as exploration and colonization of the New World. Campbell et al. (1999) specify the event that marked the company's insertion in society as the entity responsible for it, which occurred in 1953, with the case of AP Smith Manufacturing Company, whose shareholders opposed the donation of financial resources to Princeton University. In this case, the interpretation of the Supreme Court of New Jersey, regarding the incorporation of the corporation into society, was in favor of donating resources to Princeton University, contrary to the interests of the group of shareholders. This resulted in the creation of the corporate philanthropy law, which would determine that a corporation could contribute to social and cultural development (Campbell et al., 1999). Carroll (1999) also expounds his point of view on the origin of CSR, stating that the first concerns with the theme of social responsibility emerged during the 30s and 40s, with the most important references of this period being the works of Barnard (The Functions of the Executive, 1938), Clark (Social Control of Business, 1939) and Kreps (Measurement of the Social Performance of Business, 1940). Wood (1991) identifies four important historical moments to understand the process of formation of the concept of CSR: feudalism, mercantilism, and the periods of industrialization and post-industrialization. During feudalism, the social commitment was with God, the Church, and the people in general. With the advent of mercantilism, the economic commitment in Europe became the expansion and strengthening of the nation-state, above all through the intensification of colonial expeditions. In the industrialization phase, the focus shifted to increasing production as a means of expanding capital and ensuring the return on investment for owners. If, on the one hand, this strategy resulted in a strong increase in mass production and improved the population's standard of living, on the other, the absence of a social policy contributed to the exploitation of work and the natural environment. The fourth moment, called post-industrialization, is marked by a scenario of discontinuity. Economic growth is seen as the main instrument of social progress and companies are beginning to be labeled for their immoral exploitation practices and the difficulty in solving specific social needs. It is now believed that organizations should be able not only to maintain the level of wealth, but also to assume their social responsibilities.

Barbier (1987) states that the emergence of CSR is linked to a new paradigm called eco-development. For the author, the most immediate concern of this new paradigm is the free play of market forces and the pressures arising from the externalities produced, capable of breaking the balance of the Environment on a world scale. And it is in this juncture of ideas that issues such as CSR arise and gain relevance, so that the actions of governments and organizations can be questioned. Indeed, many authors have identified this link between CSR and the Environment (Dresse et al., 2021; Ullah & Nasim, 2021; Unsworth et al., 2021).

Thus, there are several theories about the origin of CSR, anticipating that we are in the presence of an old paradigm that has persisted over time, as it proves advantageous for companies that embrace it in their management practices. That's why it is important to go through some of the reasons that lead to the involvement of organizations with this concept.

2.2 *The Reasons of Corporate Social Responsibility*

We have already noted that the nature of CSR can be substantive (Enderle & Tavis, 1998; Jones, 1996; Knautz, 1997; Kok et al., 2001; Mintzberg et al., 2002; Mulligan, 1986; Wood, 1991) or instrumental (Barry, 2002; Drucker, 1984; Friedman, 1962; Levitt, 1958; Marrewijk, 2003; McWilliams & Siegel, 2001; Stroup & Neubert, 1987;

Swanson, 1995; Zwetsloot, 2003), that is, it may be based on ethical interests or may aim at achieving purely economic ends. According to several studies, we can state that the reasons that lead companies to engage in CSR practices are due to demands from interested parties, such as investors (Costello, 2021; Mackey et al., 2007), NGOs, and Government (Donaldson & Preston, 1995; Hoang et al., 2021; Hu et al., 2021; Wood, 1991), consumers (Costello, 2021; Kim & Ji, 2021; McWilliams & Siegel, 2001; Overton et al., 2021; Schuler & Cording, 2006; van Doorn et al., 2021; Vătămănescu et al., 2021), and workers, local community and wider community (Aguilera et al., 2006; Ali et al., 2021; Barnett, 2007; Baron, 2001; Chang, Lin, et al., 2021; Chang, Hu, et al., 2021; Costello, 2021; Guo et al., 2021; Hassanie et al., 2021; Kang et al., 2021; Puchalska-Kamińska et al., 2021; Wood, 1991; Wu et al., 2021; Yin et al., 2021).

In a study conducted by Gunningham et al. (2004), it is mentioned that multinational companies are highly vulnerable to pressure from stakeholders, such as consumers, activists, and shareholders, which leads them to engage in CSR practices in partnership with NGOs. For van de Ven and Graafland (2006), there are two reasons that lead companies to develop CSR practices: strategic reasons and moral reasons. Strategic reasons indicate that the company is involved in CSR to maximize its profitability. Moral motives include community-rooted values that motivate managers to "do the right thing" (Harwood et al., 2011; Kang & Wood, 1995; Sachdeva & Panfil, 2008). The reasons for the company's involvement in CSR actions also mentioned in the study by Garay and Font (2012), in addition to reasons of an altruistic nature, are also reasons inherent to competitiveness (brand differentiation and efficiency) and to compliance with stakeholders. Hemingway and Maclagan (2004) state that the reasons that lead a company to get involved in CSR can be associated with the personality of the company's leader, given that a manager decides to plan a CSR action based on his own personal values. Some authors (Aaronson, 2005; Fox et al., 2002; Moon, 2004) have analyzed the role and influence of Government as a driver of CSR. On the other hand, the internationalization of companies, such as the installation of subsidiaries, consortia, and partnerships, drive the transfer of some CSR practices in the form of a corporate governance system (Brammer et al., 2012; Tengblad & Ohlsson, 2010). Marrewijk (2003) identified the following motivations: compliance with legislation, improving the image and the importance of environmental sustainability and the involvement of organizations at this level.

Also, regarding the fact that it is advantageous to develop CSR actions, we can refer to Carroll's opinion, who points out that this type of behavior is, in the long term, in the company's own interest. This author argues that if the company needs to have a healthy environment to exist in future, it must act now, and this can ensure its long-term viability (Carroll, 1999). In the same line of thought, Davis and Blomstrom (1975) suggest that it is better to deal with social problems now than having to deal with them later, which would take up the company's time and could thus hinder the achievement of its production goals. In addition, according to Davis and Blomstrom (1975) and Frederick et al. (1988), the socially responsible behavior of companies provides a balance between the power and responsibility of the company, so that it recognizes its moral obligations and thus try to correct the social (and environmental) problems caused by its activities. In a way, this can discourage government regulation, as well as improve the company's image, promoting long-term profit. Following this idea, CSR is seen as a consequence of the pressures imposed by stakeholders on companies, which, in order to avoid greater government regulation or obtain concessions, are forced to take administrative decisions that do not necessarily contribute to their profit, but that somehow can help them in future (Varadarajan & Menon, 1988).

Undoubtedly, one of the stakeholders most cited as one of those that most contributes to the positive performance of companies from socially responsible programs are employees (Barreira & Machado, 2019). There are countless authors who refer to the good performance of employees when they feel motivated by the social practices of the companies where they work (Ackfeldt & Wong, 2006; Ashford & Mael, 1989; Ashmos & Duchon, 2000; Baumeister & Leary, 1995; Burroughs & Eby, 1998; Cameron et al., 2004; Christopher et al., 2004; Dutton et al., 1994; Eisenberger et al., 2001; Gavin & Mason, 2004; Gouldner, 1960; Haller & Hadler, 2006; Herrbach & Mignonac, 2004; Kernbach & Schutte, 2005; Kets de Vries, 2001; Kets de Vries & Florent-Treacy's, 2002; Koys, 2001; Kriger & Hanson, 1999; Milliman et al., 2003; Sergeant & Frenkel, 2000; Settoon et al., 1996; Shemwell et al., 1998; Wright & Cropanzano, 2004). CSR policies can go even further when companies are interested in reconciling family life with the professional life of their workers (Faria & Machado, 2018). According to Hay et al., by getting involved in social issues, the publicity generated "tends to improve the company's public image, which can result in direct economic benefits" but which, according to the authors, "are, however, difficult to measure" (Hay et al., 1976, p. 12). Therefore, strategic social responsibility or "strategic philanthropy" (Saiia et al., 2003; Wulfson, 2001) helps to achieve the strategic objectives of the business, as good actions are beneficial both for business and for society in general (Lantos, 2001). In this way, the study of philanthropic actions that improve the strategic position of companies has been the object of analysis by several authors, including Porter and Kramer (2002) and Saiia et al. (2003). Kanter (1999) reinforces this idea, stating that there are companies that use the concept of CSR as an instrument of corporate social innovation, exemplifying with Bell Atlantic, IBM, the Marriott International group, United Airlines, and BankBoston, which get involved on social issues strategically to innovate organizational processes and products. Burke and Logsdon (1996) also consider the issue from an instrumental point of view and state that CSR pays for itself, as it generates substantial benefits to the company's activities. Therefore, CSR is a strategic tool, capable of helping the company to achieve its mission and sustain its core business (Burke & Logsdon, 1996). Other authors reinforce the positive relationship between the social and financial performance of companies such as Preston and O'Bannon (1997), Waddock and Graves (1997), Tiras et al. (1998), and Statman (2000).

Indeed, there are several studies that demonstrate the positive effect that CSR has on business. When conducting a survey of 51 articles, in which they analyzed the relationship between CSR and financial performance, Griffin and Mahon (1997) highlighted three important issues: the focus on samples from different industries, the multiple dimensions of social responsibility and the need for several measures to assess the financial performance of companies that invest in CSR. Later, Ogden and

Watson (1999) studied the case of a company, relating the improvement in customer service with shareholder profits and highlighted that, despite the company incurring higher costs, the financial returns for shareholders improved. Ruf et al. (2001) also investigated the relationship between social and economic performance, concluding that changes in social performance were positively associated with sales growth. Joyner and Payne (2002) mention a survey of employees of companies that they themselves considered ethical, in which the degree of commitment was extremely high.

Reference should also be made to an important study conducted by Orlitzky et al. (2003), which points to a positive correlation between CSR and corporate financial performance (PFC), in both directions. In other words, CSR can serve as a catalyst for the PFC, and the opposite can also occur. Following this line of research, Hillman and Keim (2001) studied several companies, from the point of view of their social performance, and concluded that there was a greater creation of shareholder value in the cases of companies that followed a management based on the suggested principles stakeholder management than those that did not have them as a priority. Aupperle et al. (1985) describe several works that associate CSR with profit, since the 1970s.

Indeed, the relationship between company performance and CSR is a subject that has been the subject of several studies (Chang, Lin, et al., 2021; Falkenberg & Brunsæl, 2011; Gonçalves et al., 2021; Gupta & Krishnamurti, 2021; Jan et al., 2021; Kuzey et al., 2021; Thuy et al., 2021; Ying et al., 2021).

Wood (1991) considered that the principles that motivate a company's social responsibility actions are divided into three levels of analysis: institutional, organizational, and individual. Thus, the motivation for CSR actions can result from the principle of legitimacy (institutional level), that is, through the desire to maintain credibility and legitimacy as a socially responsible actor in a shared environment. Alternatively, motivation may stem from an organizational principle of public responsibility, particularly for results related to the primary and secondary areas of the company. Finally, motivation can result from the individual choices of managers, their personal preferences for responsibility, and their inclinations.

Thus, there are many factors that motivate companies to invest in CSR. But what are the main issues that guide companies' choices today, portrayed by a pandemic crisis scenario, in which companies must take key decisions, namely in terms of their social responsibilities?

3 Corporate Social Responsibility in Times of Crisis—A Brief Analysis Considering the COVID-19 Pandemic

During the COVID-19 pandemic, many companies were affected. As this is a social crisis, naturally companies were not immune, because they are made up of people

(Kucharska & Kowalczyk, 2019). In this context, the most recent public health awareness campaigns, by companies, have promoted social commitment and increased CSR (Bapuji et al., 2020).

In fact, the growing pressure from civil society for companies to act as socially responsible organizations is not recent, so companies have been expanding their CSR strategies to adapt them to a more dynamic world (Droppert & Bennett, 2015). In this way, they have been responding to several epidemiological and demographic changes (Droppert & Bennett, 2015).

According to Mahmud et al. (2021), there have been several CSR contributions in critical periods. Some examples are: the 2004 Asian Tsunami (Chong, 2009; Fernando, 2010), the 2005 Hurricane Katrina in the United States (Johnson et al., 2011; Zhao et al., 2015), the 2006 Central Java Earthquake in Indonesia (Hendarto, 2009), the 2008 Hurricane Ike in the United States (Wang, 2009), the 2008 Weather Damaged Crop in the United States (Johnson et al., 2011), the 2008 Sichuan Earthquake in China (Zhao et al., 2015), the 2010 Haiti Earthquake (Madsen & Rodgers, 2014), the 2011 Great East Japan Earthquake and Tsunami (Kouadio et al., 2012), and the 2013 collapse of Rana Plaza in Bangladesh (Begum & Solaiman, 2016; Hasan et al., 2017). The same happened with the 2020 COVID-19 pandemic (Bapuji et al., 2020; Guan et al., 2020).

However, the data that allow us to determine to what extent the economic and social recovery at a global level is associated with the organizations' CSR actions is still scarce. Although, we can refer to a study developed by Mahmud et al. (2021), which sought to investigate the philanthropic initiatives of some organizations in the US, to protect their employees, customers, and communities, during the period of the COVID-19 pandemic.

Regarding employees, we can say that their involvement in CSR actions brings several benefits to organizations (Xiao et al., 2020) and is an opportunity for them to get involved in social projects, which often have the purpose of solving social crises (Bode & Singh, 2018). Given the current context, the effect of job loss is one of the main challenges, so it is important for employees to count on the support of companies and on the philanthropic efforts of co-workers (Bapuji et al., 2020). Therefore, at a critical period such as the present, employers and employees must build an even stronger relationship than before. Companies must adopt protection policies for their workers. In the same vein, employees must react positively by engaging with the implementation of companies' CSR policies.

Regarding customers, they pay a significant level of attention to social crises and demand that companies not only operate following the logic of profit, but also contribute to reducing the adverse effects of numerous crises that have arisen in several ways, such as natural disasters, financial crises, epidemic/pandemics, and human failures (Boccia & Sarnacchiaro, 2020). CSR scholars report several positive impacts of CSR initiatives for socially responsible companies, including increased customer satisfaction, consumer loyalty, company identification, consumer trust, corporate and brand reputation, brand preference, and purchase intentions (Randle et al., 2019). CSR policies also allow consumers to identify with companies that adopt those actions (Baskentli et al., 2019). Edinger-Schons et al. (2019) state that customers who feel a strong bond with a company will tend to support it.

Regarding communities, we can say that CSR is a strategic management tool that offers perspectives of mutual gain for the community and for the company (Boadi et al., 2019; Mahmud et al., 2020). Community pressure influences the company's CSR policy and its implementation strategies (Skouloudis et al., 2015). The community provides companies with social licenses to operate (Boadi et al., 2019), if they generate various benefits through their social initiatives, to offset the adverse effects of their operations (Mahmud et al., 2020; Randle et al., 2019).

The COVID-19 pandemic has heightened communities' perceptions of social inequalities (Bapuji et al., 2020). Thus, the trust that is established between companies and the communities where they operate is essential to satisfy mutual interests (Boadi et al., 2019), with CSR playing a fundamental role here, because social problems can and should be solved based on the idea of mutual aid (Aknin & Whillans, 2020).

Within the scope of the aforementioned study (Mahmud et al., 2021), we can see that several companies were involved, at the level of their employees, in CSR actions related to remote work, cleaning conventions and hygiene protocols, use of safety equipment, social distancing practices, travel restrictions, in addition to the offer of paid leave, help programs in case of illness and health care benefits, bonuses, benefits associated with employee volunteering and employee assistance programs. These internal CSR initiatives demonstrate that several employers have shown respect for their employees during this critical period (Deng et al., 2020).

Mahmud et al. (2021) also state that many companies were concerned and committed to meeting the needs of their customers, namely through the food supply, medicines, protective supplies, medical equipment, information, and other services.

Some companies have also launched several initiatives addressed to families and communities, such as "BringHomethe Fun," from Hasbro. Many others have carried out donation policies in collaboration with non-profit organizations. These external CSR practices have shaped the relationships between companies and customers and communities during this crisis promoted by COVID-19 (Deng et al., 2020), having demonstrated that organizations and society are interconnected (McLennan & Banks, 2019).

In another study carried out in Greece (Panagiotopoulos, 2021), a new type of CSR called critical CSR was identified, which is situated in a hybrid position between tactical and strategic CSR. CSR's response to the current pandemic appears tactical in that it is unexpected, flexible, short term, and quick. At the same time, it seems strategic, as it is robust in the resources used and powerful in building trust and social capital. Also in this study, several urgent CSR actions were identified by organizations that tried to protect their employees by applying remote work, less business travels, less physical meetings, greater use of information technology resources and training actions on protective measures against COVID-19. Companies also acted externally, offering money, medical equipment and services to hospitals, research foundations, and civil protection organizations, among others. Indeed, it is possible to state that the pandemic was transformed into an opportunity for large corporations to understand more deeply the role they must play at the level of the CSR, which must have a strategic role, with long-term goals. The response given by companies demonstrates that CSR during crises can resist and even grow stronger.

Thus, it is important to re-examine the role of companies for social well-being at different times, particularly at a time of crisis. Either way, during an acute crisis, a company can improve its business reputation and relationships with employees, customers, and communities. By assuming themselves as socially responsible entities, companies can claim their power in society (Davis & Blomstrom, 1975).

In short, today, and more than ever, companies must seek not only their financial performance, but also the benefit of society and the well-being of their stakeholders (Mahmud et al., 2021).

4 Conclusion

In conclusion, we can see that the literature—based on practical studies—has been consensual about the benefits of CSR in organizations (even from an instrumental point of view), helping them not only to be more productive and efficient, but also to assume a more responsible and friendly position regarding their stakeholders.

The current situation has placed enormous pressure on organizations, but at the same time it has created a favorable scenario for investing in more sustainable and socially responsible strategies. Indeed, these strategies involve thinking globally, but acting locally, and establishing synergies with several partners. And it is interesting to observe that these partners are not always the obvious ones, but the ones that, at a given moment, make more sense. Seeing a car factory collaborating with a hospital, to supply ventilators, reflects this new context.

And it was precisely this type of interaction that the pandemic has accustomed us to. In fact, many companies took advantage of this critical situation to explore the opportunity to promote CSR practices, whether out of necessity, pressure from their stakeholders or willingly. The reason that has mobilized companies in this direction may not always be the most genuine and desirable, but the fact is that we have witnessed, all over the world, to an increase in the involvement of companies with good social practices. The current challenge is to maintain this trend and integrate it into the companies' strategy, in a more consistent and sustainable way.

Here too, government action can be of great relevance, to encourage companies in this recovery process, in a more sustainable manner and without leaving out their stakeholders.

So, CSR can be the most genuine link between companies and society. It can be a way for organizations to adapt to the post-pandemic world. It can be a means to an end. It can be the focus of recovery and resilience plans. It can be the contribution of organizations to economic, social, and environmental recovery. It can be an opportunity for companies and industries to innovate and progress in a sustainable way and in line with the interests and needs of its stakeholders.

Perhaps in the next crisis that the world will experience, if companies are more aligned with their stakeholders, at the CSR level, they will be able to provide a more effective response to the problems that may arise.

Bibliography

Aaronson, S. A. (2005). "Minding our business": What the United States government has done and can do to ensure that U.S. Multinationals act responsibly in foreign markets. *Journal of Business Ethics, 59*(1), 175–198.

Ackfeldt, A., & Wong, V. (2006). The antecedents of prosocial service behaviors: An empirical investigation. *The Services Industries Journal, 26*(7), 727–745.

Aguilera, R. V., Williams, C. A., Conley, J. M., & Rupp, D. E. (2006). Corporate governance and social responsibility: A comparative analysis of the UK and the US. *Corporate Governance: An International Review, 14*(3), 147–158.

Aknin, L. B., & Whillans, A. V. (2020). Helping and happiness: A review and guide for public policy. *Social Issues and Policy Review*, 1–32. Advance online publication.

Ali, M., Islam, T., Mahmood, K., Ali, F. H., & Raza, B. (2021). Corporate social responsibility and work engagement: Mediating roles of compassion and psychological ownership. *Asia-Pacific Social Science Review, 21*(3), 196–213.

Altman, B. (1998). Corporate community relations in the 1990s: A study in transformation. *Business and Society, 37*(2), 221–227.

Ashford, B. E., & Mael, F. (1989). Social identity theory and the organization. *Academy of Management Review, 14*(1), 20–39.

Ashmos, D. P., & Duchon, D. (2000). Spirituality at work: A conceptualization and measure. *Journal of Management Inquiry, 9*(2), 134–145.

Aupperle, K. D., Carroll, A. B., & Hatfield, J. D. (1985). An empirical examination of the relationship between corporate social responsibility and profitability. *Academy of Management Journal, 28*(2), 446–463.

Bapuji, H., Patel, C., Ertug, G., & Allen, D. G. (2020). Corona crisis and inequality: Why management research needs a societal turn. *Journal of Management, 46*, 1205–1222.

Barbier, E. (1987). The concept of sustainable development. *Environmental Conservation, 14*(2), 101–110.

Barnard, C. (1938). *The functions of the executive*. Harvard University Press.

Barnett, M. L. (2007). Stakeholder influence capacity and the variability of financial returns to corporate social responsibility. *Academy of Management Review, 32*(3), 794–816.

Baron, D. P. (2001). Private politics, corporate social responsibility, and integrated strategy. *Journal of Economics & Management Strategy, 10*(1), 7–45.

Barreira, A. F., & Machado, C. F. (2019). Transversal competences: A true and effective support to achieve greater organizational sustainability. In C. Machado (Ed.), *Sustainable human resource management: Policies and practices*. River Publishers.

Barry, N. (2002). The stakeholder concept of corporate control is illogical and impractical. *The Independent Review, 6*(4), 541–554.

Baskentli, S., Sen, S., Du, S., & Bhattacharya, C. B. (2019). Consumer reactions to corporate social responsibility: The role of CSR domains. *Journal of Business Research, 95*, 502–513.

Baumeister, R. F., & Leary, M. R. (1995). The need to belong: Desire for interpersonal attachments as a fundamental human motivation. *Psychological Bulletin, 117*(3), 497–529.

Begum, A., & Solaiman, S. M. (2016). Rana disaster: How far can we proceed with CSR? *Journal of Financial Crime, 23*(4), 748–768.

Bendheim, C. L., Waddock, S. A., & Graves, S. B. (1998). Determining best practice in corporate-stakeholder relations using data envelopment analysis. *Business and Society, 37*(3), 302–338.

Berman, S. L., et al. (1999). Does stakeholder orientation matter? The relationships between stakeholder management models and firm financial performance. *Academy of Management Journal, 42*, 488–506.
Boadi, E. A., He, Z., Bosompem, J., Say, J., & Boadi, E. K. (2019). Let the talk count: Attributes of stakeholder engagement, trust, perceive environmental protection and CSR. *SAGE Open, 9*(1), 1–15.
Boccia, F., & Sarnacchiaro, P. (2020). Chi-squared automatic interaction detector analysis on a choice experiment: An evaluation of responsible initiatives on consumers' purchasing behavior. *Corporate Social Responsibility and Environmental Management, 27*(2), 1143–1151.
Bode, C., & Singh, J. (2018). Taking a hit to save the world? Employee participation in a corporate social initiative. *Strategic Management Journal, 39*, 1003–1030.
Bowen, H. (1953). *Social responsibility of the businessman*. Harper and Row.
Brammer, S., Jackson, G., & Matten, D. (2012). Corporate social responsibility and institutional theory: New perspectives on private governance. *Socio-Economic Review, 10*(1), 3–28.
Burke, L., & Logsdon, J. M. (1996). How corporate social responsibility pays off. *Long Range Planning, 29*(4), 495–502.
Burroughs, S. M., & Eby, L. T. (1998). Psychological sense of community at work: A measurement system and explanatory framework. *Journal of Community Psychology, 26*(6), 509–532.
Cameron, K. S., Bright, D., & Caza, A. (2004). Exploring the relationships between organizational virtuousness and performance. *The American Behavioral Scientist, 47*(6), 766–790.
Campbell, L., Gulas, C. S., & Gruca, T. S. (1999). Corporate giving behavior and decision-maker social consciousness. *Journal of Business Ethics, 19*(4), 375–383.
Carroll, A. B. (1999). Corporate social responsibility: Evolution of a definitional construct. *Business and Society, 38*(3), 268–295.
Carroll, A. B., & Schwartz, M. (2003). Corporate social responsibility: A three-domain approach. *Business Ethics Quarterly, 13*(4), 503–530.
Chang, C.-H., Lin, H.-W., Tsai, W.-H., Wang, W.-L., & Huang, C.-T. (2021). Employee satisfaction, corporate social responsibility, and financial performance. *Sustainability, 13*(18), 2–25.
Chang, Y.-P., Hu, H.-H., & Lin, C.-M. (2021). Consistency or hypocrisy? The impact of internal corporate social responsibility on employee behavior: A moderated mediation model. *Sustainability, 13*(17), 2–21.
Chong, M. (2009). Employee participation in CSR and corporate identity: Insights from a disaster-response program in the Asia-Pacific. *Corporate Reputation Review, 12*(2), 106–119.
Christopher, A. N., et al. (2004). Materialism and affective well-being: The role of social support. *Personality and Individual Differences, 37*(3), 463–470.
Clark, J. M. (1939). *Social control of business*. McGraw-Hill.
Costello, K. W. (2021). What is the social responsibility of companies? *Electricity Journal, 34*(8), 107008.
Davis, K., & Blomstrom, R. L. (1975). *Business and society: Environment and responsibility*. McGraw-Hill.
Deng, X., Long, X., Schuler, D. A., Luo, H., & Zhao, X. (2020). External corporate social responsibility and labor productivity: A S-curve relationship and the moderating role of internal CSR and government subsidy. *Corporate Social Responsibility and Environmental Management, 27*(1), 393–408.
Donaldson, T., & Preston, L. (1995). The Stakeholder theory of the corporation: Concepts, evidence, and implications. *Academy of Management Review, 20*(1), 65–91.
Dresse, A., Nielsen, J. Ø., & Fischhendler, I. (2021). From corporate social responsibility to environmental peacebuilding: The case of bauxite mining in Guinea. *Resources Policy, 74*, 1873–7641.
Droppert, H., & Bennett, S. (2015). Corporate social responsibility in global health: An exploratory study of multinational pharmaceutical firms. *Globalization and Health, 11*(15), 1–8.
Drucker, P. F. (1984). The new meaning of corporate social responsibility. *California Management Review, 26*(2), 53–63.

Dutton, J. E., Duberich, J. M., & Harquail, C. V. (1994). Organizational images and member identification. *Administrative Science Quarterly, 39*(2), 239–263.
Edinger-Schons, L. M., Lengler-Graiff, L., Scheidler, S., & Wieseke, J. (2019). Frontline employees as corporate social responsibility (CSR) ambassadors: A quasi-field experiment. *Journal of Business Ethics, 157*, 359–373.
Eisenberger, R., et al. (2001). Reciprocation of perceived organizational support. *Journal of Applied Psychology, 86*(1), 42–51.
Enderle, G., & Tavis, L. A. (1998). A balanced concept of the firm and the measurement of its long-term planning and performance. *Journal of Business Ethics, 17*(11), 1129–1144.
Evan, W. M., & Freeman, R. E. (1993). A stakeholder theory of the modern corporation: Kantian capitalism. In T. L. Beauchamp & N. E. Bowie (Eds.), *Ethical theory and business* (pp. 75–84). Prentice-Hall.
Falkenberg, J., & Brunsæl, P. (2011). Corporate social responsibility: A strategic advantage or a strategic necessity? *Journal of Business Ethics, 99*, 9–16.
Faria, A., & Machado, C. F. (2018). Work-family conciliation policies: Answering to corporate social responsibility—A case study. In C. F. Machado & P. J. Davim (Eds.), *Corporate social responsibility in management and engineering*. River Publishers.
Fernando, M. (2010). Corporate social responsibility in the wake of the Asian tsunami: Effect of time on the genuineness of CSR initiatives. *European Management Journal, 28*(1), 68–79.
Fox, T., Ward, H., & Howard, B. (2002). *Public sector roles in strengthening corporate social responsibility: A baseline study strengthening corporate social responsibility*. World Bank.
Frederick, W. (1986). Toward CSR3—Why ethical analysis is indispensable and unavoidable in corporate affairs. *California Management Review, 28*(2), 126–141.
Frederick, W. (1994). From CSR1 to CSR2. *Business and Society., 33*(2), 150–164.
Frederick, W. (1998). Moving to CSR4: What to pack for the trip. *Business and Society, 37*(1), 40–59.
Frederick, W., Davis, K., & Post, J. (1988). *Business and society: Corporate strategy, public policy*. McGraw-Hill.
Freeman, E. (1984). *Strategic management: A stakeholder approach*. Pitman Publishing.
Friedman, M. (1962). *Capitalism and freedom*. The University Chicago Press.
Fry, L. W., Gerald, D. K., & Meiners, R. E. (1982). Corporate contributions: Altruistic or for-profit? *Academy of Management Journal, 25*(1), 96–104.
Garay, L., & Font, X. (2012). Doing good to do well? Corporate social responsibility reasons, practices and impacts in small and medium accommodation enterprises. *International Journal of Hospitality Management, 31*(2), 329–337.
Garriga, E., & Melé, D. (2004). Corporate social responsibility theories: Mapping the territory. *Journal of Business Ethics, 53*(12), 51–71.
Gavin, J. H., & Mason, R. O. (2004). The virtuous organization: The value of happiness in the workplace. *Organizational Dynamics, 33*(4), 379–392.
Gonçalves, T., Gaio, C., & Ferro, A. (2021). Corporate social responsibility and earnings management: Moderating impact of economic cycles and financial performance. *Sustainability, 13*(17), 2–14.
González-Masip, J. (2021). The concept of corporate social responsibility: Study perspectives proposal. In G. Martín-de Castro & J. González-Masip (Eds.), *Knowledge management for corporate social responsibility* (pp. 1–17). IGI Global.
Gouldner, A. W. (1960). The norm of reciprocity: A preliminary statement. *American Sociological Review, 25*(2), 161–178.
Griffin, J., & Mahon, J. (1997). The corporate social performance and the corporate financial performance debate. *Business and Society, 36*(1), 5–27.
Guan, D., et al. (2020). Global supply-chain effects of COVID-19 control measures. *Nature Human Behaviour, 4*, 577–5587.
Gunningham, N., Kagan, R. A., & Thornton, D. (2004). Social licence and environmental protection: Why businesses go beyond compliance. *Law and Social Inquiry, 29*(2), 307–341.

Guo, M., Ahmad, N., Adnan, M., Khalil-Ur-Rehman, & Naveed, R. T. (2021). The relationship of CSR and employee creativity in the hotel sector: The mediating role of job autonomy. *Sustainability, 13*(17), 2–14.

Gupta, K., & Krishnamurti, C. (2021). Corporate social responsibility, competition, and firm value. *Pacific Basin Finance Journal, 68*, 101622.

Haller, M., & Hadler, M. (2006). How social relations and structures can produce happiness and unhappiness: An international comparative analysis. *Social Indicators Research, 75*(2), 169–216.

Harrison, J. S., & Freeman, R. E. (1999). Stakeholders, social responsibility, and performance: Empirical evidence and theoretical perspectives. *Academy of Management Journal, 42*(5), 479–485.

Harwood, I., Humby, S., & Harwood, A. (2011). On the resilience of corporate social responsibility. *European Management Journal, 29*(4), 283–290.

Hasan, M. M., Mahmud, A., & Islam, M. S. (2017). Deadly incidents in Bangladeshi apparel industry and illustrating the causes and effects of these incidents. *Journal of Finance and Accounting, 5*(5), 193–199.

Hassanie, S., Karadaş, G., & Lawrence Emeagwali, O. (2021). Do CSR perceptions influence work outcomes in the health care sector? The mediating role of organizational identification and employee attachment. *Sustainability, 13*(17), 2–16.

Hay, R., Gray, E. R., & Gates, J. E. (1976). *Business and society*. Southwestern Publishing.

Hemingway, C. A., & Maclagan, P. W. (2004). Managers' personal values as drivers of corporate social responsibility. *Journal of Business Ethics, 50*(1), 33–44.

Hemphill, T. (2004). Monitoring global corporate citizenship: Industry self-regulation at a crossroads. *Journal of Corporate Citizenship, 14*, 81–95.

Hendarto, K. A. (2009). The implementation of corporate social responsibility (CSR) in Central Java earthquake: A preliminary study on consumer belief, attitude, and purchase intention. *Gadjah Mada International Journal of Business, 11*(3), 409–441.

Herrbach, O., & Mignonac, K. (2004). How organizational image affects employee attitudes. *Human Resource Management Journal, 14*(4), 76–88.

Hillman, A. J., & Keim, G. D. (2001). Shareholder value, stakeholder management, and social issues: What's the bottom line? *Strategic Management Journal, 22*(2), 125–139.

Hoang, T. M. N., Nguyen, T. T. T., & Phan, T. T. H. (2021). Government environmental regulation, corporate social responsibility, ecosystem innovation strategy and sustainable development of Vietnamese seafood enterprises. *International Journal of Data and Network Science, 5*(4), 713–726.

Hood, J. (1998). Do corporations have social responsibilities? *The Freeman, 48*(11), 680–684.

Hu, M., Xiong, W., & Xu, C. (2021). Analyst coverage, corporate social responsibility, and firm value: Evidence from China. *Global Finance Journal, 50*, 100671.

Hummels, H. (1998). Organizing ethics: A stakeholder debate. *Journal of Business Ethics, 17*(13), 1403–1419.

Hutchins, M. J., & Sutherland, J. W. (2008). An exploration of measures of social sustainability and their application to supply chain decisions. *Journal of Cleaner Production, 16*(15), 1688–1698.

Jan, N., Karn, A. L., Li, Z., & Liu, X. (2021). The relationship between corporate social responsibility reporting and firm performance and the moderating role of firm life cycle stages: Evidence from China. *Sustainability, 13*(18), 2–12.

Johnson, B. R., Connolly, E., & Carter, T. S. (2011). Corporate social responsibility: The role of Fortune 100 companies in domestic and international natural disasters. *Corporate Social Responsibility and Environmental Management, 18*(6), 352–369.

Jones, T. (1996). Missing the forest for the trees: A critique of the social Responsibility concept and discourse. *Business e Society, 35*(1), 7–41.

Joyner, B., & Payne, D. (2002). Evolution and implementation: A study of values, business ethics and corporate social responsibility. *Journal of Business Ethics, 41*(4), 297–311.

Kang, S.-E., Lee, C.-K., Moon, Y.-J., Park, Y.-N., & Suess, C. (2021). Impact of CSR on organizational behavior during a pandemic: Highlighting public health and safety in the airline industry. *Sustainability, 13*(17), 2–17.

Kang, Y. C., & Wood, D. J. (1995). *Before-profit social responsibility-turning the economic paradigm upside-down*. Proceedings of the 6th Annual Meeting of International Association for Business and Society, Vienna, 408–418.

Kanter, R. M. (1999). From spare change to real change. *Harvard Business Review, 77*(3), 122–132.

Keinert, C. (2008). *Corporate social responsibility as an international strategy*. Physica-Verlag, A Springer Company.

Kernbach, S., & Schutte, N. S. (2005). The impact of service provider emotional intelligence on customer satisfaction. *Journal of Services Marketing, 19*(7), 438–444.

Kets de Vries, M. R. (2001). Creating authentizotic organizations: Well-functioning individuals in vibrant companies. *Human Relations, 54*(1), 101–111.

Kets de Vries, M. R., & Florent-Treacy, E. (2002). Creating high commitment organizations. *Organizational Dynamics, 30*(4), 295–309.

Kim, S., & Ji, Y. (2021). Positive ripple effects of corporate leaders' CSR donations amid COVID-19 on corporate and country reputations: Multi-level reputational benefits of CSR focusing on Bill Gates and Jack Ma. *Public Relations Review, 47*(4), 102073.

Knautz, R. (Ed.). (1997). *Corporate social responsibility*. Policy Spotlight.

Kok, P., Van der Weile, T., McKenna, R., & Brown, A. (2001). A corporate social responsibility audit within a quality management framework. *Journal of Business Ethics, 31*(4), 285–297.

Kouadio, I. K., Aljunid, S., Kamigaki, T., Hammad, K., & Oshitani, H. (2012). Infectious diseases following natural disasters: Prevention and control measures. *Expert Review of Anti-Infective Therapy, 10*(1), 95–104.

Koys, D. (2001). The effects of employee satisfaction, organizational citizenship behavior and turnover on organizational effectiveness: A unit-level, longitudinal study. *Personnel Psychology, 54*(1), 101–114.

Kreps, T. J. (1940). Measurement of the social performance of business. In *An investigation of concentration of economic power for the temporary national economic committee* (Monograph nº7). Washington, DC: US Government Printing Office.

Kriger, M. P., & Hanson, B. J. (1999). A value-based paradigm or creating truly healthy organizations. *Journal of Organizational Change Management, 12*(4), 302–317.

Kucharska, W., & Kowalczyk, R. (2019). How to achieve sustainability? Employee's point of view on company's culture and CSR practice. *Corporate Social Responsibility and Environmental Management, 26*(2), 453–467.

Kuzey, C., Uyar, A., Nizaeva, M., & Karaman, A. S. (2021). CSR performance and firm performance in the tourism, healthcare, and financial sectors: Do metrics and CSR committees' matter? *Journal of Cleaner Production, 319*, 128802.

Lantos, G. (2001). The boundaries of strategic corporate social responsibility. *Journal of Consumer Marketing, 18*(7), 595–630.

Levitt, T. (1958). The dangers of social responsibility. *Harvard Business Review, 36*(5), 41–50.

Logsdon, J. M., & Yuthas, K. (1997). Corporate social performance, stakeholder orientation, and organizational moral development. *Journal of Business Ethics, 16*(12–13), 1213–1226.

Mackey, A., Mackey, T. B., & Barney, J. B. (2007). Corporate social responsibility and firm performance: Investor preferences and corporate strategies. *Academy of Management Review, 32*(3), 817–835.

Madsen, P. M., & Rodgers, Z. J. (2014). Looking good by doing good: The antecedents and consequences of stakeholder attention to corporate disaster relief. *Strategic Management Journal, 36*, 776–794.

Magalhães, C. (2011). Responsabilidade social e regulada no sector metalomecânico brasileiro e português: um estudo comparativo e cultural. In N. C. Maria Alice et al (Orgs.), *Responsabilidade Social: Uma visão ibero-americana* (pp. 81–109). Almedina.

Mahmud, A., Ding, D., Kiani, A., & Hasan, M. (2020). Corporate social responsibility programs and community perceptions of societal progress in Bangladesh: A multimethod approach. *SAGE Open, 10*(2), 1–17.
Mahmud, A., Ding, D., Kiani, A., & Hasan, M. (2021). Corporate social responsibility: Business responses to coronavirus (COVID-19) pandemic. *SAGE Open, 11*(1), 1–17.
Maignan, I., & Ferrell, O. C. (2000). Measuring corporate citizenship in two countries: The case of the United States and France. *Journal of Business Ethics, 23*(3), 283–297.
Maignan, I., & Ferrell, O. C. (2001a). Antecedents and benefits of corporate citizenship: An investigation of French businesses. *Journal of Business Research, 51*(1), 37–51.
Maignan, I., & Ferrell, O. C. (2001b). Corporate citizenship as a marketing instrument: Concepts, evidence and research directions. *European Journal of Marketing, 35*(3–4), 457–484.
Marrewijk, M. V. (2003). Concepts and definitions of CSR and corporate sustainability: Between agency and communion. *Journal of Business Ethics, 44*(2–3), 95–105.
Matten, D., Crane, A., & Chapple, W. (2003). Behind the mask: Revealing the true face of corporate citizenship. *Journal of Business Ethics, 45*(1/2), 109–120.
McGuire, J. W. (1963). *Business and society*. McGraw-Hill.
McLennan, S., & Banks, G. (2019). Reversing the lens: Why corporate social responsibility is not community development. *Corporate Social Responsibility and Environmental Management, 26*(1), 117–126.
McWilliams, A., & Siegel, D. (2001). Corporate social responsibility: A theory of the firm perspective. *Academy of Management Review, 26*(1), 117–127.
Milliman, J., Czaplewski, A. J., & Ferguson, J. (2003). Workplace spirituality and employee work attitudes: An exploratory empirical assessment. *Journal of Organizational Change Management, 16*(4), 426–447.
Mintzberg, H., Simons, R., & Kunal, B. (2002). Beyond selfishness. *MIT Sloan Management Review, 44*(1), 67–74.
Mitchel, R. K., Agle, B. R., & Wood, D. J. (1997). Toward a theory of stakeholder identification and salience: Defining the principle of who and what really counts. *Academy of Management Review, 22*(4), 853–886.
Moon, J. (2004). Government as a driver of corporate social responsibility. *ICCSR Research Paper Series, 44*(20), 1–27.
Mulligan, T. (1986). A critique of Milton Friedman's essay "The social responsibility of business is to increase its profits." *Journal of Business Ethics, 5*, 265–269.
Ogden, S., & Watson, R. (1999). Corporate performance and stakeholder management: Balancing shareholder and customer interests in the UK privatized water industry. *Academy of Management Journal, 42*(5), 526–538.
Orlitzky, M., et al. (2003). Corporate social and financial performance: A meta-analysis. *Organization Studies, 24*(3), 403–441.
Overton, H., Kim, J. K., Zhang, N., & Huang, S. (2021). Examining consumer attitudes toward CSR and CSA messages. *Public Relations Review, 47*(4), 102095.
Panagiotopoulos, I. (2021). Novel CSR & novel coronavirus: Corporate social responsibility inside the frame of coronavirus pandemic in Greece. *International Journal of Corporate Social Responsibility, 6*(10).
Peterson, D. K. (2004). The relationship between perceptions of corporate citizenship and organizational commitment. *Business and Society, 43*(3), 296–319.
Phillips, R., et al. (2003). What stakeholder theory is not. *Business Ethics Quarterly, 13*(4), 479–502.
Porter, M. E., & Kramer, M. R. (2002). The competitive advantage of corporate philanthropy. *Harvard Business Review, 80*(12), 56–68.
Preston, L. E., & O'Bannon, D. P. (1997). The corporate social-financial performance relationship: A typology and analysis. *Business and Society, 36*(4), 419–429.
Puchalska-Kamińska, M., Łądka-Barańska, A., & Roczniewska, M. (2021). Social purpose in an organization from the perspective of an employee: A self-determination outlook on the meaning of work. *BMC Research Notes, 14*(1), 1–6.

Randle, M., Kemperman, A., & Dolnicar, S. (2019). Making cause-related corporate social responsibility (CSR) count in holiday accommodation choice. *Tourism Management, 75*, 66–77.

Ruf, B. M., et al. (2001). An empirical investigation of the relationship between change in corporate social performance and financial performance: A stakeholder theory perspective. *Journal of Business Ethics, 32*(2), 143–156.

Sachdeva, A., & Panfil, O. (2008). *CSR perceptions and activities of small and medium enterprises (SMEs) in seven geographical clusters: Survey report*. UNIDO.

Saiia, D., Carroll, A. B., & Buchholtz, A. (2003). Philanthropy as strategy. *Business e Society, 42*(2), 169–201.

Schuler, D. A., & Cording, M. (2006). A corporate social performance—Corporate financial performance behavioral model for consumers. *Academy of Management Review, 31*(3), 540–588.

Sergeant, A., & Frenkel, S. (2000). When do customer contact employees satisfy customers? *Journal of Service Research, 3*(1), 18–34.

Sethi, S. P. (1975). Dimensions of corporate social responsibility. *California Management Review, 17*(3), 58–64.

Settoon, R. P., Bennett, N., & Liden, R. C. (1996). Social exchange in organizations: Perceived organizational support, leader-member exchange, and employee reciprocity. *Journal of Applied Psychology, 81*(2), 219–227.

Shemwell, D. J., Yavas, U., & Bilgin, Z. (1998). Customer-service provider relationships: An empirical test of a model of service quality, satisfaction and relationship-oriented outcomes. *International Journal of Service Industry Management, 9*(2), 155–168.

Skouloudis, et al. (2015). Priorities and perceptions of corporate social responsibility. *Management Decision, 53*(2), 375–401.

Statman, M. (2000). Socially responsible mutual funds. *Financial Analysts Journal, 56*(3), 30–39.

Stoney, C., & Winstanley, D. (2001). Stakeholding: Confusion or utopia? Mapping the conceptual terrain. *Journal of Management Studies, 38*(5), 603–626.

Stroup, M. A., & Neubert, R. L. (1987). The evolution of social responsibility. *Business Horizons, 30*, 22–24.

Swanson, D. L. (1995). Addressing a theoretical problem by reorienting the corporate social performance model. *Academy of Management Review, 20*(1), 43–64.

Tengblad, S., & Ohlsson, C. (2010). The framing of corporate social responsibility and the globalization of national business systems: A longitudinal case study. *Journal of Business Ethics, 93*(4), 653–669.

Thuy, C. T. M., Khuong, N. V., Canh, N. T., & Liem, N. T. (2021). Corporate social responsibility disclosure and financial performance: The mediating role of financial statement comparability. *Sustainability, 13*(18), 2–14.

Tiras, S., Ruff, B., & Brown, R. M. (1998). *The relation between stakeholders' implicit claims and firm value.*

Ullah, S., & Nasim, A. (2021). Do firm-level sustainability targets drive environmental innovation? Insights from BRICS economies. *Journal of Environmental Management, 294*, 112754.

Unsworth, K. L., Davis, M. C., Russell, S. V., & Bretter, C. (2021). Employee green behaviour: How organizations can help the environment. *Current Opinion in Psychology, 42*, 1–6.

Varadarajan, P. R., & Menon, A. (1988). Cause-related marketing: A coalignment of marketing strategy and corporate philanthropy. *Journal of Marketing, 52*(3), 58–74.

van de Ven, B., & Graafland, J. J. (2006). Strategic and moral motivation for corporate social responsibility. *Journal of Corporate Citizenship, 22*, 111–123.

van Doorn, J., Risselada, H., & Verhoef, P. C. (2021). Does sustainability sell? The impact of sustainability claims on the success of national brands' new product introductions. *Lavoro e Diritto, 35*(2), 77–104.

Vătămănescu, E.-M., Dabija, D.-C., Gazzola, P., Cegarro-Navarro, J. G., & Buzzi, T. (2021). Before and after the outbreak of Covid-19: Linking fashion companies' corporate social responsibility approach to consumers' demand for sustainable products. *Journal of Cleaner Production, 321*, 128945.

Vos, J. (2003). Corporate social responsibility and the identification of stakeholders. *Corporate Social Responsibility and Environmental Management, 10*, 141–152.
Waddock, S., & Graves, S. B. (1997). The corporate social performance-financial performance link. *Strategic Management Journal, 18*(4), 303–319.
Wang, X. (2009). *Corporate social responsibility practices in responding to natural disasters: The United States and China*. University of Florida.
Warhurst, A. (2001). Corporate citizenship and corporate social investment. *Journal of Corporate Citizenship, 1*, 57–73.
Windsor, D. (2019). Defining corporate social responsibility for developing and developed countries: Comparing proposed approaches. In Information Resources Management Association. *Corporate social responsibility: Concepts, methodologies, tools, and applications* (pp. 1–27). IGI Global.
Wood, D. J. (1991). Corporate social performance revisited. *Academy of Management Review, 16*(4), 691–718.
Wood, J. D., & Logsdon, J. M. (2002). Business citizenship: From domestic to global level of analysis. *Business Ethics Quarterly, 12*(2), 155–187.
Wright, T. A., & Cropanzano, R. (2004). The role of psychological well-being in job performance: A fresh look at an age-old quest. *Organizational Dynamics, 33*(4), 338–351.
Wu, Q., et al. (2021). The role of CSR and ethical leadership to shape employees' pro-environmental behavior in the era of industry 4.0. A case of the banking sector. *Sustainability, 13*(17), 2–20.
Wulfson, M. (2001). The ethics of corporate social responsibility and philanthropic ventures. *Journal of Business Ethics, 29*, 135–145.
Xiao, M., Cooke, F. L., Xu, J., & Bian, H. (2020). To what extent is corporate social responsibility part of human resource management in the Chinese context? A review of literature and future research directions. *Human Resource Management Review, 30*(4), Article 100726.
Yin, C., Ma, H., Gong, Y., Chen, Q., & Zhang, Y. (2021). Environmental CSR and environmental citizenship behavior: The role of employees' environmental passion and empathy. *Journal of Cleaner Production, 320*, 128751.
Ying, M., Tikuye, G. A., & Shan, H. (2021). Impacts of firm performance on corporate social responsibility practices: The mediation role of corporate governance in Ethiopia corporate business. *Sustainability, 13*(17), 2–21.
Zhao, M., Wang, F., Zhao, D., & Wei, J. (2015). Using CSR theory to examine disaster aid response to the Wenchuan earthquake. *Global Business and Organizational Excellence, 34*, 50–59.
Zwetsloot, G. (2003). From management systems to corporate social responsibility. *Journal of Business Ethics, 44*(2/3), 201–207.

Interviewer Characteristics and Nonresponse Survey Outcomes: A Portuguese Case Study

Cláudia Cunha, Alice Delerue Matos, Gina Voss, and Carolina Machado

Abstract Unit and item nonresponse are a source of great concern as they are capable of lowering the quality of survey data. Although the literature is unanimous in highlighting interviewers as important elements capable of influencing survey outcomes, there are still several inconsistencies regarding which interviewer characteristics might influence unit and item nonresponse, as well as few studies focusing on the role of intrapersonal skills. To understand if different interviewer characteristics are at work producing unit and item nonresponse rates and to identify those characteristics, this study uses data from the Survey of Health, Aging and Retirement in Europe (SHARE), more precisely from the Interviewer Survey and the Wave 6 Questionnaire (Household Income module). More specifically, multiple linear regression analysis was performed to assess the association between nonresponse survey outcomes with sociodemographic characteristics, experience and intrapersonal skills. Results show that better unit response rates are mostly associated with interviewers' intrapersonal skills, such as persistency and resilience. Furthermore, better income response rates are associated with positive expectations and self-confidence. These findings have implications for recruitment, selection and training decisions.

Keywords Interviewers · Unit nonresponse · Income nonresponse · Intrapersonal skills · SHARE

C. Cunha (✉) · G. Voss
Communication and Society Research Centre, Institute of Social Sciences, University of Minho, Braga, Portugal
e-mail: claudiacunha@ics.uminho.pt

A. D. Matos
Department of Sociology, Institute of Social Sciences, University of Minho, Braga, Portugal

C. Machado
Department of Management, School of Economics and Management, Interdisciplinary Centre of Social Sciences (CICS.NOVA.UMinho), University of Minho, Braga, Portugal

C. Machado (ed.), *Challenges and Trends in Organizational Management and Industry*, Management and Industrial Engineering,
https://doi.org/10.1007/978-3-030-98048-1_5

1 Introduction

The quality of survey data can be severely jeopardized by two growing threats, namely the decline in cooperation by sample members, referred to as unit nonresponse and the lack of answers to questions within an interview, that is, item nonresponse (Groves & Couper, 1998; Singer & Ye, 2013; Tourangeau & Yan, 2007).

According to Burton et al. (2006), in the British Household Survey, about 75% of respondents in wave 4 had responded to all previous waves. In wave 13, this percentage had dropped to 52.7%, with refusals being indicated as the main reason for this fall. In addition, a study by Tourangeau and Yan (2007) revealed that, in some studies, the item nonresponse rate may range from 20 to 40% for questions related to financial aspects and income.

In all interviewer-mediated surveys, interviewers are highlighted as an important determinant of both unit and item nonresponse outcomes (Pickery & Loosveldt, 1998). Even though efforts have been made to explain interviewer effects such as unit and item nonresponse, it remains unclear which interviewer characteristics play a role in survey nonresponse. Besides, fewer studies portray the role of intrapersonal skills on survey nonresponse, when compared to sociodemographic characteristics and experience.

Therefore, this study aims to understand the characteristics of the Portuguese interviewers working for the Survey of Health Ageing and Retirement in Europe (SHARE) that are associated with the unit and item nonresponse rates.

A key strength of this study lies in the unique matching of several interviewer features available in the SHARE Interviewer Survey with the interviewers' unit and item nonresponse rates. Thus, this research makes it possible not only to determine if different mechanisms are at work producing unit and item nonresponse rates but also to identify those mechanisms, which is critical for recruitment, selection and training decisions.

2 Interviewers and Nonresponse Survey Outcomes

Several studies conclude that the sociodemographic characteristics of interviewers do not tend to have substantial effects on the unit and item nonresponse rates (Groves & Couper, 1998; Lemay & Durand, 2002; Morton-Williams, 1993; Pickery & Loosveldt, 1998; Vassallo et al., 2015; West & Blom, 2017). However, there are some exceptions, with researches pointing out that gender, age and education are capable of influencing both cooperation (Blom et al., 2011; Fowler & Mangione, 1990; Hansen, 2006; Haunberger, 2010; Jäckle et al., 2013; Singer et al., 1983) and the number of unanswered questions (Berk & Bernstein, 1988; Riphahn & Serfling, 2005; Schräpler, 2006; Singer et al., 1983).

Even though not all studies were able to explain why these sociodemographic characteristics influence survey outcomes, some attempts were made using the principle of liking. Based on Petty and Cacioppo (1986) and Chaiken's (1980) works, Groves et al. (1992) conclude that the decision to participate in a study is mainly processed through heuristic mechanisms. This means that the individuals do not think about the possibility of participation diligently and the cooperation is often based on some trait of the interviewers or the organization they represent. According to the authors, people using a heuristic decision approach can participate through different principles, such as the principle of liking, in which potential sample members cooperate because they like or share similarities with the interviewer. Some authors found evidence of this principle, suggesting that liking a specific sociodemographic trait of the interviewer or sharing similar sociodemographic characteristics, increases the probability of cooperation (Durrant et al., 2010), and response to financial questions (Vercruyssen et al., 2017).

A much less explored line of work within the subject of sociodemographic characteristics is the interviewers' employment status. As reported by Blohm et al. (2006), the interviewers who work full time at another job besides interviewing have more difficulty contacting and convincing people to cooperate in a study. The authors explain these results through the shorter amount of time interviewers have available to find and persuade sample members.

Experience is another relevant interviewer trait commonly associated with different interviewer effects, even though studies about its consequences on response rates are rather mixed. One line of thought is that the experience acquired with different surveys and measurements will positively influence both unit (Jäckle et al., 2013) and item response rates (Essig & Winter, 2009). Nevertheless, the mechanisms underlying these positive relationships are still weakly understood (West & Blom, 2017). According to Groves and Couper (1998), more experienced interviewers acquire skills that make them better at tailoring their approach to the type of situation at the doorstep and at maintaining interaction with sample members, which positively influences cooperation rates. Furthermore, experience with a certain survey and its questions might boost interviewers' confidence which, in turn, may shape their behaviour while interacting with the respondent and positively affect response rates (Blom & Korbmacher, 2013).

However, experience is not always portrayed as positive for surveys. Singer et al (1983) concluded that, while having some experience is more beneficial than having none, the more experienced an interviewer, the more their unit response rate declines. Similarly, Tu and Liao (2007) found that more experienced interviewers got more refusals and "don't know" answers to questions related to behaviours. These findings may result from the assignment of difficult cases to more experienced interviewers (Blom et al., 2011) and heavy workloads (Pickery et al., 2001; Singer et al., 1983).

Another more recent strand of research focuses on the effects of interviewers' soft skills on response rates. Klaus (2010) highlights that only 25% of professional success derive from technical knowledge and the remaining 75% are dependent on soft skills. This also applies to the interviewer job, since during contact and cooperation and sometimes even during the application of the questionnaire, interviewers will have to make use of their intrapersonal and interpersonal skills to achieve good results.

The study of the effects of intrapersonal skills on unit response rates first started with the work of Lehtonen (1996), that developed an interviewer attitudes scale and concluded that persuasion is positively correlated with cooperation. More recent studies point out that interviewers who are confident about their abilities have higher cooperation rates (Durrant & D'Arrigo, 2014). Durrant et al. (2010) also conclude that lower refusal rates are found among interviewers who adopt a positive attitude towards reluctance. Furthermore, using the 2013 German refresher sample for SHARE, Olson et al. (2020) found that individuals are more likely to cooperate when interviewers report lower acceptance of respondents' refusals when they agree that the voluntary nature of studies should be emphasized and when they report that most people will participate if contacted at the right time.

It is believed that persistence, self-confidence and positivity have positive effects on unit response rates because interviewers with these traits adopt behaviours and strategies that differ from the ones adopted by interviewers who do not possess these skills. For instance, Campanelli, Sturgis and Purdon (1997) showed that interviewer persistence in doing follow-up calls reduced noncontact rates.

Singer and Kohnke-Aguirre (1979) also show that the item nonresponse rate is influenced by negative interviewers' expectations regarding the respondents' answers to survey questions, especially those of a sensitive nature. Similar results were found in a more recent study where interviewers with positive expectations regarding the attainment of income answers achieved a lower income nonresponse rate, whereas those who were more pessimistic reached worse outcomes (Friedel, 2020). According to Wuyts and Loosveldt (2017), expectations can influence survey outcomes because interviewers, consciously or unconsciously, tend to act in accordance with their own expectations. Therefore, interviewers with lower expectations might not neutrally ask questions or might be unable to probe correctly.

The unclear findings regarding the identification of the interviewer features that might influence unit and item nonresponse and the few studies portraying the role of intrapersonal skills on survey nonresponse indicate the need for further research into this subject.

3 Data and Methods

3.1 Data

SHARE is a European multidisciplinary and cross-national study that provides internationally comparable longitudinal microdata on health, socioeconomic status and social and family networks of more than 140,000 individuals aged 50 and over, covering 28 European countries and Israel. To understand the changes experienced by SHARE respondents over time, every two years the same people are interviewed at their homes, using the computer-assisted personal interviewing (CAPI) technique. Further methodological details of the SHARE project are available elsewhere (Börsch-Supan et al., 2013; Malter & Börsch-Supan, 2017).

With the purpose of developing new strategies to preserve or even increase the quality of the survey data, the central SHARE coordination team (SHARE Central) implemented the SHARE Interviewer Survey (Blom & Korbmacher, 2013). Through this survey, SHARE Central intended to acquire a deeper understanding of how interviewers influence survey outcomes. In wave 6, SHARE interviewers from 12 countries (Portugal, Austria, Belgium, Germany, Spain, Sweden, Slovenia, Estonia, Greece, Italy, Luxembourg and Poland) participated in this survey. The interviewer questionnaire is answered after the national training sessions, but before beginning fieldwork and the data is collected in an online survey. The data from the Interviewer Survey can be linked to the SHARE survey data through an interviewer's (scrambled) identification code.

Even though the Interviewer Survey was developed to explain interviewer effects, such as unit and item nonresponse, the SHARE Interviewer Survey dataset does not include these interviewer performance indicators. Nevertheless, it is possible to calculate them for each interviewer. SHARE data allows all registered users to assess the interviewer item nonresponse but only the national country team/survey agency can calculate the unit nonresponse rate. In fact, the results of the interviewer's attempts to conduct the interviews are only available in the export file of the Case CTRL application used to administer contact attempts and conduct interviews.

For the current research, the Portuguese SHARE team calculated the unit nonresponse rate for each interviewer of the country. Nevertheless, for reasons of confidentiality aimed at preserving the anonymity of responses to the Interviewer Survey, the national country team/survey agency is not able to associate this interviewer's performance indicator to the Interviewer's Survey data, where each interviewer is identified by a scrambled identification code. Access to the key used to scramble the interviewer identification code is only available to SHARE Central which, in turn, does not have access to the personal identification of the interviewers. For this reason, the unit nonresponse rates, calculated by the national country team, were later sent to SHARE Central, which added them to the Portuguese Interviewer Survey dataset.

This study uses data from the SHARE Interviewer Survey (release 6.0.0) and the Wave 6 Survey Questionnaire—Household Income module (HH), which enables access to the income nonresponse. As unit nonresponse can only be calculated for our own country, the sample is restricted to Portuguese interviewers ($N = 39$).

In Portugal, the unit nonresponse rate for wave 6, at the individual level, was 17.0% (Malter & Börsch-Supan, 2017) and the income nonresponse rate was, on average, 32.2% ($SD = 29.9$).

3.2 Variables

3.2.1 Independent Variables

Based on the literature review, this study includes the following sociodemographic variables: age at the time of the Interviewer Survey (2015); gender (0 = male and 1 = female); education (1 = lower-level secondary school; 2 = medium-level secondary school; 3 = upper-level secondary school; 4 = university degree); full-time employed (0 = No and 1 = Yes); part-time employed (0 = No and 1 = Yes); unemployed (0 = No and 1 = Yes); student (0 = No and 1 = Yes).

The experience of interviewers is analyzed through the variable years working as an interviewer (calculated by reference to the Interviewer Survey).

Intrapersonal skills were assessed using the following question about interviewers' attitudes towards persuasion: "*Now follow a series of statements on persuading respondents. Interviewers may differ in their opinions about these strategies. There is no right or wrong answer. We are interested in your opinion, based on your experience as an interviewer*". Therefore, the interviewers are presented with a list of eight situations, with which they have to strongly disagree, somewhat disagree, somewhat agree or strongly agree. Some arguments stress that all reluctant respondents must be persuaded to participate, that with enough effort, even the most reluctant respondent can be persuaded to participate and that, when approached at the right time, the majority of respondents will cooperate. Conversely, other statements highlight that participant refusals must be accepted, that it makes no sense to contact reluctant people repeatedly and that respondents who agree to participate after a great deal of effort will not provide reliable answers. Other sentences also underline the respect for the respondent's privacy as well as the voluntary nature of the participation. Even though these items have been employed to derive summed attitude scores (Blohm et al., 2006; De Leeuw et al., 1998; Hox & De Leeuw, 2002; Olson et al., 2020), in this study the attitude items were used as separate variables as a result of the low Cronbach's alpha coefficient (0.051).

The interviewers' expectations about obtaining answers to the income question were also considered in the analysis of the characteristics underlying income nonresponse: "*Social surveys very often ask about respondents' income. What do you expect, how many of your respondents (in percentage) in SHARE will provide information about their income?*".

3.2.2 Dependent Variables

To determine the unit nonresponse rates, the contact history of 1350 Portuguese respondents was analyzed. Thus, for each interviewer, the unit nonresponse rate was calculated as follows:

$$\text{Unit nonresponse rate} = \frac{\text{Number of failed interviews}}{\text{Number of eligible respondents}} \times 100$$

This formula makes it possible to obtain information about the percentage of individuals who could not be interviewed in the universe of eligible people assigned to each interviewer. Eligible respondents are those whom the interviewers visited or tried to visit, regardless of whether they were able to establish any interaction. The number of failed interviews include the following situations: no answer, no contact (when interviewers visit sample members but are unable to establish any interactions); contact, try again (when sample members ask interviewers to come back another day, but the interview was not conducted); contact, appointment for interview (when there is an appointment for an interview, but the interview never took place); contact, appointment for another contact (when a specific day was scheduled to address the possibility of cooperation, but the interview never took place); soft refusals; hard refusals; deceased (when a respondent is deceased but the end-of-life interview with a proxy respondent is not conducted); in a care home (when respondents in care homes are not interviewed); moved—new address known (when a respondent moved to a new address known to the interviewer, but the interview did not take place).

The income nonresponse rate was measured using the household income question (hh017) in the HH module of the respondent questionnaire: "*How much was the overall income, after taxes and contributions, that your entire household had in an average month in 2014?*". Answering the income question was coded as 0 and not answering this question was coded as 1.

3.3 Methods

This study was developed in two stages, and all analyses were performed using the software R (4.0.5) and the software Statistical Package for the Social Sciences (SPSS, 25.0).

As a first step, univariate descriptive statistics were conducted, in order to characterize our study population.

As a second step, multiple linear regression analysis was performed to assess the effect of interviewer characteristics (sociodemographic characteristics, experience, attitudes and expectations) on the unit and income nonresponse rates. Due to the small sample size ($N = 39$), this analysis was performed using the bootstrap technique. The bootstrap is a statistical method that provides more accuracy to statistical estimates when the sample size is small (Efron & Tibshirani, 1993). In the bootstrap analysis, 2000 samples of the same size as the original sample were drawn, with replacement from the original dataset.

All non-significant variables were removed from the models using a backward stepwise selection process. Therefore, even though all the independent variables cited above were tested, only significant results are shown in the tables.

Additionally, the beta value (B), standard error (SE) and significance (where p-values of < 0.05 were considered statistically significant) are presented in the tables below.

4 Results

Table 1 shows the sociodemographic characteristics and experience of the study population.

Portuguese interviewers are, on average, 35.4 (SD = 8.1) years old and the vast majority are women (71.8%). In addition, more than half of the interviewers have a university degree (82.1%) and 17.9% have upper-level secondary education. Concerning their employment situation, 20.5% of the Portuguese interviewers are full-time employed, 17.9% are part-time employed, 33.3% are unemployed and 28.2% are students.

Concerning experience, the participants have been working as interviewers for 6.1 years (SD = 7.1).

Table 1 Sociodemographic characteristics and experience of the interviewers

	Portuguese interviewers
Sociodemographic characteristics	
Age, *mean (SD)*	35.4 (8.1)
Gender *(%)*	
Male	28.2
Female	71.8
Education *(%)*	
Upper-level secondary school	17.9
University degree	82.1
Full-time employed *(%)*	
No	79.5
Yes	20.5
Part-time employed *(%)*	
No	82.1
Yes	17.9
Unemployed *(%)*	
No	66.7
Yes	33.3
Student *(%)*	
No	71.8
Yes	28.2
Experience	
Years working as an interviewer, *mean (SD)*	6.1 (7.1)

Source SHARE Interviewer Survey, wave 6, release 6.0.0; Unweighted data, N (unweighted) = 39

The attitudes towards persuasion and expectations of the interviewers are displayed in Table 2.

Most interviewers agree that reluctant respondents should always be persuaded to participate (92.3%) and that with enough effort, even the most reluctant respondent can be persuaded to cooperate in a study (92.3%). Moreover, 97.4% of interviewers agree that the privacy of the respondents should be respected. With respect to accepting refusals in the event of reluctance, more than half of the Portuguese interviewers showed their disagreement towards this approach (59%). Descriptive statistics also reveal that the vast majority of interviewers agree that the voluntary nature of the study should always be emphasized (94.9%). Moreover, 56.4% of Portuguese interviewers disagree that it does not make sense to contact reluctant target persons repeatedly. Concerning the respondents' willingness to participate if contacted at the right time, all Portuguese interviewers support this statement. Besides, 77.1% of interviewers are opposed to the argument that respondents who are persuaded after a great effort do not provide reliable answers.

Furthermore, the interviewers expect that, on average, 79.2% (SD = 14.6) of their respondents will provide an answer to the income question.

Table 3 presents the multiple linear regression findings for the association between different interviewer characteristics and the unit nonresponse rate.

The results show that there is a significant negative relationship between being employed full time and the unit nonresponse rate ($B = -28.242$; $SE = 9.047$; $p = 0.004$). Similar results were found for the interviewers who reported being part-time employed ($B = -54.32$; $SE = 8.901$; $p < 0.001$). Thus, lower unit nonresponse rates are associated with being employed full time and part-time in another job, in addition to the interviewer duties. No other sociodemographic characteristics or experience were significantly associated with unit nonresponse.

The results also indicate that some attitudes towards persuasion are associated with unit nonresponse. More specifically, there is a positive association between agreeing that a refusal should be accepted in the event of reluctance and the unit nonresponse rate ($B = 9.116$; $SE = 4.383$; $p = 0.046$). Therefore, being more accepting of respondents' refusals is shown to be associated with poor interviewer performance. Moreover, the interviewers who agree with the argument "Respondents who were persuaded after great effort do not provide reliable answers" achieve better survey outcomes ($B = -9.974$; $SE = 4.794$; $p = 0.042$).

The results of the multiple linear regression for the income nonresponse rates (Table 4) revealed that no interviewers' sociodemographic characteristics or experience were associated with income nonresponse. Likewise, no significant associations were found between the interviewers' attitudes towards persuasion and income nonresponse rates.

Concerning to interviewers' expectations, statistical analysis shows that the income nonresponse rate decreases 0.757% for each additional percentage point on the expected income ($B = -0.757$; $SE = 0.290$; $p = 0.002$). Therefore, higher and more positive expectations regarding the attainment of income answers are associated with a lower lack of responses to the income question and, thereupon, with a better interviewer performance.

Table 2 Attitudes towards persuasion and expectations of the interviewers

	Portuguese interviewers
Attitudes towards persuasion	
Persuasion to participate (%)	
Somewhat disagree	7.7
Somewhat agree	35.9
Strongly agree	56.4
Enough effort (%)	
Somewhat disagree	7.7
Somewhat agree	25.6
Strongly agree	66.7
Respect the privacy of the respondent (%)	
Strongly disagree	2.6
Somewhat disagree	15.4
Strongly agree	82.1
Accept refusal (%)	
Strongly disagree	20.5
Somewhat disagree	38.5
Somewhat agree	38.5
Strongly agree	2.6
Emphasize that participation is voluntary (%)	
Strongly disagree	2.6
Somewhat disagree	2.6
Somewhat agree	28.2
Strongly agree	66.7
Not contact reluctant target persons repeatedly (%)	
Strongly disagree	28.2
Somewhat disagree	28.2
Somewhat agree	35.9
Strongly agree	7.7
Right time (%)	
Somewhat disagree	26.3
Strongly agree	73.7
Unreliable answers (%)	
Strongly disagree	42.9
Somewhat disagree	34.3
Somewhat agree	20.0
Strongly agree	2.9
Expectations	

(continued)

Table 2 (continued)

	Portuguese interviewers
Expected percentage of answers to the income question, *mean (SD)*	79.2 (14.6)

Source SHARE Interviewer Survey, wave 6, release 6.0.0; Unweighted data, N (unweighted) = 39

Table 3 Multiple linear regression for unit nonresponse

	B	**SE**	*p*
Intercept	100.828	31.848	0.003
Sociodemographic characteristics			
Full-time employed (yes)	−28.242	9.047	0.004
Part-time employed (yes)	−54.32	8.901	< 0.001
Attitudes towards persuasion			
Accept refusal (agree)	9.116	4.383	0.046
Emphasize that participation is voluntary (agree)	−12.323	7.333	0.096
Unreliable answers (agree)	−9.974	4.794	0.042

Source SHARE Interviewer Survey, wave 6, release 6.0.0

Table 4 Multiple linear regression for income nonresponse

	B	**SE**	*p*
Intercept	106.751	25.627	0.043
Expectations			
Expected percentage of answers to the income question	−0.757	0.290	0.002
Attitudes towards persuasion			
Persuasion to participate (agree)	−24.685	7.078	0.053

Source SHARE Interviewer Survey, wave 6, release 6.0.0

5 Discussion

Interviewers play a key role in the quality of any interviewer-mediated survey since they are responsible for several tasks during the entire data collection process. These tasks include achieving the cooperation of sample members and getting answers to questions during the interview. However, not all interviewers are successful at persuading sample members to participate and at obtaining the requested information in a survey, entailing negative consequences for the quality of studies. Despite research on the interviewer traits that might influence unit and item nonresponse, the results are thus far inconclusive, with fewer studies focusing on intrapersonal characteristics when compared to sociodemographic features and experience.

Therefore, to assess if different interviewer traits are at work producing unit and income nonresponse rates and to identify those characteristics, this study focused on the association between these survey outcomes and interviewers' sociodemographic characteristics, experience, attitudes and expectations.

The results show that there are no significant associations between any of the performance indicators and interviewers' age, gender, education and experience. These findings are in line with previous studies showing that more and less successful interviewers did not differ in terms of these traits (Groves & Couper, 1998; Lemay & Durand, 2002; Morton-Williams, 1993; Pickery & Loosveldt, 1998; Vassallo et al., 2015; West & Blom, 2017).

Furthermore, contrary to the study by Blohm et al. (2006), the interviewers who work full time or part-time in another job besides interviewing are better at persuading sample members to cooperate. One possible explanation for this result lies in the extrinsic motivation resulting from the performance-related payment system employed in SHARE. In fact, people who are already employed and accept the challenge of becoming an interviewer may have done so for financial reasons. Since in SHARE the interviewers' pay and other bonuses are contingent on the number of interviews they complete, these interviewers might have felt more motivated to put in additional effort to contact and persuade sample members.

Another hypothesis that may help explain the better results obtained by people engaged in another profession is the traits of the interviewer job. More specifically, individuals who are already employed in another job may have also sought work as interviewers because of specific job aspects they might consider personally rewarding, such as being involved in scientific research or gaining insight into other peoples' social circumstances. Ergo, these aspects may operate as incentives, intrinsically motivating interviewers to achieve better survey outcomes.

The results regarding interviewers' expectations and attitudes towards persuasion revealed that different mechanisms are at play producing unit and income nonresponse.

By not agreeing to immediately accept a refusal, the interviewers proved to be persistent and resilient, traits that were revealed to be important for obtaining higher unit response rates. These findings are in line with previous research that stressed the importance of persistence for cooperation (Durrant & D'Arrigo, 2014; Durrant et al., 2010).

As previously explained, it is believed that persistence-oriented interviewers achieve better results due to behaviours and strategies adopted during fieldwork that differ from the ones adopted by interviewers who do not possess this skill. Thus, we hypothesize that those who are more persuasive and resilient might not only make more contact attempts but might also be more likely to resort to tailoring and maintaining interaction with respondents, strategies associated with high cooperation rates (Groves & Couper, 1998).

According to Groves and Couper (1998), tailoring is the term used to describe interviewers' ability to adapt their doorstep speech and strategies to the characteristics of the sample member and social surroundings. For tailoring to be effectively applied, the interviewers must firstly be able to maintain interaction with the sample member, in order to be able to pick up clues that will help them choose more suitable strategies. However, for maintaining interaction and tailoring their approach, interviewers cannot immediately accept the sample members' refusals. Instead, they must be able to endure the refusal and be persistent to keep the conversation going and find the adequate approach for a given situation. Despite the underlying difficulties of analyzing the behaviours of interviewers during contact and the doorstep interaction, it would be interesting to determine in the future research if resilience and persistence are associated with tailoring and with the ability to maintain interaction.

Besides, the analysis of the relationship between interviewers' attitudes and unit nonresponse revealed another interesting finding: believing that particularly reluctant individuals will not answer honestly does not negatively affect interviewers' cooperation rates. This might mean that interviewers try to obtain the participation of sample members regardless of their assumptions. However, this result needs more detailed research in the future.

Positive and confident expectations regarding the percentage of answers to the income question are associated with better income response rates. These findings are in line with previous research that had already stressed the importance of positivity and self-confidence for item response (Friedel, 2020; Singer & Kohnke-Aguirre, 1979). A possible explanation for this result might lie in the interviewers' behaviours, since they might act in accordance with their expectations by probing correctly or by presenting the question in a neutral fashion, not causing mistrust in the respondent.

Our study has important implications for survey practitioners. Firstly, it is important to keep interviewers motivated, as that might be related to higher cooperation rates. Furthermore, during the recruitment and selection processes, special treatment should not be given to age, gender, education or experience. Rather than privileging these traits, special attention should be paid to candidates' intrapersonal skills, such as positivity, confidence, persistence and resilience. Concerning training, more attention should be paid to interviewers' intrapersonal skills. In addition to focusing on technical skills, such as standardized interviewing rules and how to work with the CAPI software, training sessions should also address the importance of intrapersonal skills and help interviewers develop them. Furthermore, interviewers should also be reassured about the income question, for example, by clarifying the goal of the question as well as its importance, to try to circumvent the effect that negative expectations have on income nonresponse.

6 Strengths and Limitations

To our knowledge, this is the first study analyzing the influence of several interviewer characteristics on both unit and income nonresponse rates in SHARE.

However, the findings of this research need to be interpreted within the context of the study limitations.

Firstly, as mentioned previously, the Interviewer Survey was applied after the interviewer training sessions, but before the beginning of fieldwork. Thus, it is not possible to know if the interviewers' attitudes during the interviews were similar to the ones reported in the questionnaire.

Furthermore, the potential influence of respondent characteristics on the response rates achieved by interviewers could not be considered, due to the absence of data, for instance, on the characteristics of the respondents who refused to participate in the survey.

Finally, the assessment of whether the interviewer had already interviewed the same respondents in previous waves or if they were interviewing different respondents was not accounted for in the analysis.

7 Conclusion

Our findings highlight the importance of intrapersonal skills for the response rates of interviewers and, consequently, the success of studies. In fact, the interviewers with better unit response rates are more persistent and resilient and the interviewers with better income response rates are more positive and confident of their own capabilities. Thus, this study has several potential implications for the recruitment, selection and training decisions, reinforcing the need to give intrapersonal skills a prominent role in each of these processes.

Acknowledgements The authors are grateful for the funding awarded to SHARE-Portugal by the FCT/MCTES and Calouste Gulbenkian Foundation (SHARE-Portugal, 2013-2016) and by FCT/MCTES through National Funds (PIDDAC) and the European Regional Development Fund (ERDF) through the Operational Program PORNorte and PORLisboa, AACNº 01/SAICT/2016, Application nº022209 –DATALAB (SHARE-Portugal, from 2017 to the current date).

The SHARE data collection has been funded by the European Commission, DG RTD through FP5 (QLK6-CT-2001-00360), FP6 (SHARE-I3: RII-CT-2006-062193, COMPARE: CIT5-CT-2005-028857, SHARELIFE: CIT4-CT-2006-028812), FP7 (SHARE-PREP: GA N°211909, SHARE-LEAP: GA N°227822, SHARE M4: GA N°261982, DASISH: GA N°283646) and Horizon 2020 (SHARE-DEV3: GA N°676536, SHARE-COHESION: GA N°870628, SERISS: GA N°654221, SSHOC: GA N°823782) and by DG Employment, Social Affairs & Inclusion through VS 2015/0195, VS 2016/0135, VS 2018/0285, VS 2019/0332, and VS 2020/0313. Additional funding from the German Ministry of Education and Research, the Max Planck Society for the Advancement of Science, the U.S. National Institute on Aging (U01_AG09740-13S2, P01_AG005842, P01_AG08291, P30_AG12815, R21_AG025169, Y1-AG-4553-01, IAG_BSR06-11, OGHA_04-064, HHSN271201300071C, RAG052527A) and from various national funding sources is gratefully acknowledged (see www.share-project.org).

References

Berk, M. L., & Bernstein, A. B. (1988). Interviewer characteristics and performance on a complex health survey. *Social Science Research, 17*(3), 239–251. https://doi.org/10.1016/0049-089X(88)90015-4

Blohm, M., Hox, J., & Koch, A. (2006). The influence of interviewers' contact behavior on the contact and cooperation rate in face-to-face household surveys. *International Journal of Public Opinion Research, 19*(1), 97–111.

Blom, A. G., & Julie M. Korbmacher. (2013). Measuring interviewer characteristics pertinent to social surveys: A conceptual framework. *Survey Methods: Insights from the Field*, 1–16. https://doi.org/10.13094/SMIF-2013-00001.

Blom, A. G., de Leeuw, E. D., & Hox, J. J. (2011). Interviewer effects on nonresponse in the European social survey. *Journal of Official Statistics, 27*(2), 359–377. https://doi.org/10.2139/ssrn.1663704

Börsch-Supan, A., Brandt, M., Hunkler, C., Kneip, T., Korbmacher, J., Malter, F., Schaan, B., Stuck, S., & Zuber, S. (2013). Data resource profile: The survey of health, ageing and retirement in europe (share). *International Journal of Epidemiology, 42*(4), 992–1001. https://doi.org/10.1093/ije/dyt088

Burton, J., Laurie, H., & Lynn, P. (2006). The long-term effectiveness of refusal conversion procedures on longitudinal surveys. *Journal of the Royal Statistical Society. Series A: Statistics in Society, 169*(3), 459–478. https://doi.org/10.1111/j.1467-985X.2006.00415.x.

Campanelli, P., Sturgis, P., & Purdon, S. (1997). *Can you hear me knocking? An investigation into the impact of interviewers on survey response rates*. http://natcen.ac.uk/

Chaiken, S. (1980). Heuristic versus systematic information processing and the use of source versus message cues in persuasion. *Journal of Personality and Social Psychology, 39*(5), 752–766. https://doi.org/10.1037/0022-3514.39.5.752

De Leeuw, E., Hox, J., Snijkers, G., & De Heer, W. (1998). Interviewer opinions, attitudes and strategies regarding survey participation and their effect on response. In *Nonresponse in survey research. Proceedings of eighth international workshop on household survey nonresponse 24–26 September 1997* (pp. 239–248).

Durrant, G. B., & D'Arrigo, J. (2014). Doorstep interactions and interviewer effects on the process leading to cooperation or refusal. *Sociological Methods and Research, 43*(3), 490–518. https://doi.org/10.1177/0049124114521148

Durrant, G. B., Groves, R. M., Staetsky, L., & Steele, F. (2010). Effects of interviewer attitudes and behaviors on refusal in household surveys. *Public Opinion Quarterly, 74*(1), 1–36. https://doi.org/10.1093/poq/nfp098

Efron, B., & Tibshirani, R. (1993). *An introduction to the bootstrap*. Chapman & Hall.

Essig, L., & Winter, J. K. (2009). Item non-response to financial questions in household surveys: An experimental study of interviewer and mode effects. *Fiscal Studies, 30*(3–4), 367–390. https://doi.org/10.1111/j.1475-5890.2009.00100.x

Fowler, F., & Mangione, T. (1990). *Standardized survey interviewing: Mnimizing interviewer-related error*. Sage.

Friedel, S. (2020). What they expect is what you get: The role of interviewer expectations in nonresponse to income and asset questions. *Journal of Survey Statistics and Methodology, 8*(5), 851–876. https://doi.org/10.1093/jssam/smz022

Groves, R. M., Cialdini, R. B., & Couper, M. P. (1992). Understanding the decision to participate in a survey. *Public Opinion Quarterly, 56*(4), 475. https://doi.org/10.1086/269338

Groves, R., & Couper, M. (1998). Nonresponse in household interview surveys. *Wiley*. https://doi.org/10.1080/00401706.1999.10485958

Hansen, K. M. (2006). The effects of incentives, interview length, and interviewer characteristics on response rates in a CATI study. *International Journal of Public Opinion Research, 19*(1), 112–121.

Haunberger, S. (2010). The effects of interviewer, respondent and area characteristics on cooperation in panel surveys: A multilevel approach. *Quality and Quantity, 44*(5), 957–969. https://doi.org/10.1007/s11135-009-9248-5

Hox, J. J., & De Leeuw, E. D. (2002). The influence of interviewers' attitude and behavior on household survey nonresponse: An international comparison. In *Survey Nonresponse* (pp. 103–119).

Jäckle, A., Lynn, P., Sinibaldi, J., & Tipping, S. (2013). The effect of interviewer experience, attitudes, personality and skills on respondent co-operation with face-to-face surveys. *Survey Research Methods, 7*(1), 1–15. https://doi.org/10.18148/srm/2013.v7i1.4736

Klaus, P. (2010). Communication breakdown. *California Job Journal, 28*, 1–9.

Lehtonen, R. (1996). Interviewer attitudes and unit nonresponse in two different interviewing schemes. In S. Laaksonen (Ed.), *International perspectives on nonresponse* (pp. 130–140). Proceedings of the Sixth International Workshop on Household Survey Nonresponse, Statistics Finland. https://www.researchgate.net/publication/264471565_Interviewer_Effects_on_Nonresponse_in_the_European_Social_Survey

Lemay, M., & Durand, C. (2002). The effect of interviewer attitude on survey cooperation. *Bulletin of Sociological Methodology/bulletin De Méthodologie Sociologique, 76*(1), 27–44. https://doi.org/10.1177/075910630207600102

Malter, F., & Börsch-Supan, A. (2017). SHARE wave 6: Panel innovations and collecting dried blood Spots. In F. Malter & A. Börsch-Supan (Eds.), *Munich center for the economics of aging (MEA)*. Munich Center for the Economics of Aging (MEA).

Morton-Williams, J. (1993). *Interviewer approaches*. Dartmouth.

Olson, K., Smyth, J. D., Dykema, J., Holbrook, A. L., Kreuter, F., & West, B. T. (2020). *Interviewer effects from a total survey error perspective*. Chapman and hall/CRC. https://doi.org/10.1201/9781003020219

Petty, R. E., & Cacioppo, J. T. (1986). *Communication and persuasion : Central and peripheral routes to attitude change*. 262. https://books.google.com/books/about/Communication_and_Persuasion.html?hl=pt-PT&id=dUJ-AAAAMAAJ

Pickery, J., & Loosveldt, G. (1998). The impact of respondent and interviewer characteristics on the number of "no opinion" answers: A multilevel model for count data. *Quality and Quantity, 32*(1), 31–45. https://doi.org/10.1023/A:1004268427793

Pickery, J., Loosveldt, G., & Carton, A. (2001). The effects of interviewer and respondent characteristics on response behavior in panel surveys. *Sociological Methods & Research, 29*(4), 509–523.

Riphahn, R. T., & Serfling, O. (2005). Item non-response on income and wealth questions. *Empirical Economics, 30*(2), 521–538. https://doi.org/10.1007/s00181-005-0247-7

Schräpler, J. P. (2006). Explaining income nonresponse—A case study by means of the British Household Panel Study (BHPS). *Quality and Quantity, 40*(6), 1013–1036. https://doi.org/10.1007/s11135-005-5429-z

Singer, E., & Kohnke-Aguirre, L. (1979). Interviewer expectation effects : A replication and extension. *The Public Opinion Quarterly, 43*(2), 245–260.

Singer, E., & Ye, C. (2013). The use and effects of incentives in surveys. *Annals of the American Academy of Political and Social Science, 645*(1), 112–141. https://doi.org/10.1177/0002716212458082

Singer, E., Frankel, M. R., & Glassman, M. B. (1983). The effect of interviewer characteristics and expectations on response. *Public Opinion Quarterly, 47*(1), 68. https://doi.org/10.1086/268767

Tourangeau, R., & Yan, T. (2007). Sensitive questions in surveys. *Psychological Bulletin, 133*(5), 859–883. https://doi.org/10.1037/0033-2909.133.5.859

Tu, S. H., & Liao, P. S. (2007). Social distance, respondent cooperation and item nonresponse in sex survey. *Quality and Quantity, 41*(2), 177–199. https://doi.org/10.1007/s11135-007-9088-0

Vassallo, R., Durrant, G. B., Smith, P. W. F., & Goldstein, H. (2015). Interviewer effects on non-response propensity in longitudinal surveys: A multilevel modelling approach. *Journal of the Royal Statistical Society. Series A, (Statistics in Society), 178*(1), 83. https://doi.org/10.1111/RSSA.12049.

Vercruyssen, A., Wuyts, C., & Loosveldt, G. (2017). The effect of sociodemographic (mis)match between interviewers and respondents on unit and item nonresponse in Belgium. *Social Science Research, 67*, 229–238. https://doi.org/10.1016/j.ssresearch.2017.02.007

West, B. T., & Blom, A. G. (2017). Explaining interviewer effects: A research synthesis. *Journal of Survey Statistics and Methodology, 5*(2), 175–211. https://doi.org/10.1093/JSSAM/SMW024

Wuyts, C., & Loosveldt, G. (2017). The interviewer in the respondent's shoes: What can we learn from the way interviewers answer survey questions? In *Field Methods* (Vol. 29, Issue 2, pp. 140–153). Sage. https://doi.org/10.1177/1525822X16663162.

Balanced Scorecard to Improve Human Capital Management: Case Study in a Portuguese Company

Jéssica Perestrelo and Pedro Novo Melo

Abstract Balanced Scorecard allows to assess if the organization performance is aligned with its strategy, enabling management to redirect the organization and proactively act when a deviation from the defined targets occurs. This strategy focus, coupled with its results, provides management a direct feedback loop, which enables a concise action. This project documents an implementation of the Balanced Scorecard (BSC) tool, focused on the Human Capital perspective. The research method applied was a case study, in a consultant company, constituting an action-research. This work includes a literature review regarding this field of knowledge, which presents the tool and focus on the Human Capital perspective; a presentation of the organization under analysis, which includes its policies, key performance indicators, and defined strategy; and the documentation of the multiple phases of the Balanced Scorecard implementation process in the organization. New performance indicators were proposed. It was concluded that a conflict may be present among the two organization strategic drivers and two of the HR critical results and two HR efficiency processes were highlighted.

Keywords Balanced scorecard · Human Capital · Key performance indicators · Human resource management · Case study

1 Introduction

In the competitive environment of a global economy, organizations feel the need to promote their sustainability. Some ways to promote sustainability are innovation, knowledge, and efficiency. Furthermore, the measurement of organizational performance is essential for management to be able to assess the real and current

J. Perestrelo
School of Management, Polytechnic Institute of Cávado and Ave, Barcelos, Portugal

P. N. Melo (✉)
School of Technical Short Cycles, Polytechnic Institute of Cávado and Ave, Braga, Portugal
e-mail: pmelo@ipca.pt

Interdisciplinary Centre of Social Sciences—CICS.NOVA.UMinho, Braga, Portugal

C. Machado (ed.), *Challenges and Trends in Organizational Management and Industry*, Management and Industrial Engineering,
https://doi.org/10.1007/978-3-030-98048-1_6

state of their organization. Additionally, pressure from shareholders, investors, and other stakeholders to quantitatively know the organization's current performance and strategy is constant.

To respond to this problem, the management tool designed by Kaplan and Norton in 1992, called the Balanced Scorecard, emerges. This brings together four perspectives of the organization: Financial, Customers, Internal Processes, and Learning and growth of the organization, offering an integrated view of the organization. Cumulatively, this tool seeks to ensure alignment between day-to-day actions and the organization's strategy, that is, the tool works to focus all parts of the organization on the strategy defined by top management.

The Human Capital theory places people as agents of strategy, that is, it sees people as assets of the organization, who actively work to help the organization fulfill its strategy. In addition, Human Capital management seeks a contextualized measurement, focused on action.

These two approaches, Balanced Scorecard (BSC) and Human Capital Management (HCM), have two focuses in common: (1) aligning with the organization's strategy and (2) measuring organizational performance. It is, therefore, pertinent to join both approaches and forms the Human Capital Balanced Scorecard (BSC-HC).

The interest in this topic arises from the gap felt in the definition of a strategy by some organizations, but also from the belief that Human Capital is the greatest asset of organizations. It is therefore advisable to define a strategy, communicate and align Human Capital, and measure the various processes of an organization, to analyze and understand the real situation of the organization. The goals of this investigation are to study the approach, know the organization under study, and build a Balanced Human Capital Scorecard (BSC-HC) adapted to its context.

2 Literature Review

The Balanced Scorecard (BSC) is a strategic management planning tool, capable of measuring the organization's performance from four perspectives (also called quadrants or aspects): Financial, Customers, Internal Processes, and Learning and growth of the organization. The Balanced Scorecard tool—developed by Kaplan and Norton (1992)—received great prominence, as it no longer focused only on financial results, evaluated by traditional quantitative methods, and began to incorporate measures that describe the process of creating real value.

A balanced performance measurement system, together with an aligned HR system and knowledge management, forms the foundation for creating a strategically focused organization.

In the book "*The Strategy-Focused Organization: How Balanced Scorecard Companies Thrive in the New Business Environment,*" Kaplan and Norton (2000) present examples of companies that had great weaknesses, making their poor performance visible, not only in financial terms, as well as in the position they occupied in the market. Examples of these companies are Mobil North America Marketing and Refining; CIGNA Property & Casualty Insurance; Chemical (Chase) Retail Bank.

The application of the Balanced Scorecard by these organizations allowed each of them to promote major strategic and organizational changes, with visible improvement in their performance.

Its performance was based on changes in terms of strategy, focus, and organization:

- Strategy: Organizations understood the importance of describing and communicating the organization's strategy, in an understandable way, serving as a basis for its performance.
- Focus: All decisions taken began to align the organization's resources and activities with its strategy.
- Organization: Link between the various business units and their employees.

According to Bieker (2002), the Balanced Scorecard is a tool that improves the sustainable management of an organization and was created in the 1990s, with the aim of integrating intangible assets, such as "intellectual and organizational capital," into the management system more efficiently. Intangible assets are essential to ensure the organization's competitiveness. The author argues that, to maintain the focus on the most important aspects, up to twenty indicators should be integrated in the BSC instrument. This tool allows the integration of the organization's stakeholders, allowing for corporate sustainable development.

Balanced Scorecard seeks to transfer the organizational strategy to the operational level of organizations.

In favor of the Balanced Scorecard—and its perspectives—it is defended that an organizational analysis should not be carried out independently and in isolation, since these four perspectives are linked in a cause-effect relationship (Mendes et al., 2012).

2.1 *Balanced Scorecard Perspectives*

According to Pinto (2007), if there are failures in the functioning of one of the perspectives—the result of the goals not achieved in that perspective—they will compromise the efficiency in another, or in other, perspectives, reflecting on the imbalance of the Balanced Scorecard. Este desequilíbrio irá desencadear desvios no percurso estratégico, e no cumprimento da missão e da visão da organização.

The author Wu (2012) argues that the Balanced Scorecard allows organizations to improve communication with their employees, improve their strategy with a view to continuous improvement of their performance, through increased productivity. This opinion comes from the cause-effect relationship between the four aspects of the Balanced Scorecard. The perspective of *financial performance* tries to ensure the sustainability and growth of the organization, through the analysis of relevant variables, such as return on investment. The customer perspectives include identifying potential customers, defining that customer's needs, retaining current customers, and the organization's participation in the market as a whole. The perspective of *internal processes* aims, through the analysis of the set of activities, processes, and actions carried out by the organization—to understand the organization's dynamics and the improvement in the organization's effectiveness and efficiency, improving customer satisfaction. Finally, the fourth perspective, referring to the learning and growth of the organization, intends to instill the adoption of an organizational culture that motivates employees to adopt an innovative and creative attitude, to contribute to the future success, both of the individual and also of the organization (Mendes et al., 2012).

The *learning and growth* perspective—focused in this work—is presented on the basis of the Balanced Scorecard, which establishes cause-effect relationships with the other perspectives, influencing the success of the final result from the top perspective (financial perspective regarding the private sector, and customer perspective in the case of the public sector). According to Pinto (2007), vision and strategy must be at the center of the four perspectives that make up the Balanced Scorecard, as they represent the path that the organization must follow to achieve the defined goals. However, before defining the strategy, the author defends the importance of sustaining this strategy in three pillars.

2.2 Balanced Scorecard and Strategy

The focus on strategy must be constant, because only in this way will employees be able to know the direction of the organization. This focus should be encouraged by top management in order to apply a top-down approach. For Silva (2010), strategic management serves as a guide for the entire organization, as it allows for a vision of what is being achieved, and what is expected to be achieved. This provides the cooperation and commitment of employees, and consequently the increase in performance levels. To this end, the authors Kaplan and Norton (2000) present the following principles of an organization focused on strategy:

- Translating the Strategy into Operational Terms: Involvement of all employees in the organization's strategy.
- Align the Organization to the Strategy: Definition of common themes and objectives, for the different sectors that make up the organization, so that they gather guiding synergies for the success of its mission.
- Transforming the Strategy into a Task for Everyone: All employees must know and understand the organization's strategy, so that they can conduct their daily tasks, thus being motivated to carry it out.
- Convert Strategy into Continuous Process: The strategy should be an ongoing process, in order to reflect on the opportunities and threats that may arise. It is important to have a connection between the strategic budget and the operating budget. This process should be accompanied by monthly or quarterly management meetings in order to assess the current strategy and possible improvements.

- Mobilize Change through Executive Leadership: The success of an organization depends on executive leadership, that is, the ability to make the need for change something obvious to everyone. Only in this way it is possible to guarantee a united and cohesive team, capable of achieving the expected results.

The performance of the Balanced Scorecard is characterized by the creation and verification of cause-effect relationships between perspectives, objectives, goals, and indicators. Through cause-effect relationships (which imply making choices and taking decisions), it is possible to assess whether the Balanced Scorecard is reflecting the organization's strategy, that is, whether the organization's general objectives are being achieved (Pinto, 2007). In the early years of applying the Balanced Scorecard, organizations began to realize the difficulty in understanding and communicating all cause-effect relationships. To overcome this difficulty, the strategic map emerged, allowing for the expansion and success of the Balanced Scorecard.

According to Kaplan and Norton (2000), strategic maps help organizations to visualize their strategies in a cohesive, integrated, and systematic way. And, in addition to making it easier to understand, the strategic maps build the foundations of the management system, for the implementation of the strategy efficiently and quickly.

The Balanced Scorecard must reflect the organization's strategy. Its graphical representation—through strategic maps—makes it possible to communicate the results sought by the organization, and how these can be achieved. They also provide an understanding—by all organizational units and their teams—of the adopted strategy, contributing to its execution. According to Smith (2007), the strategic map identifies the main strategic themes of the organization, and how they are interconnected. The strategic map allows you to articulate the organization's strategy.

2.3 Human Capital

The famous quote by Alfred Marshall (1982) states that *"The most valuable of all capital is that invested in human beings"*. This author was one of the fathers of the awareness of human resources as a capital (an asset) of organizations. This fact highlights the need to incorporate HR into the organization's strategy.

The theory of Human Capital, presented by Ehrenberg and Smith (1997), "*conceptualizes workers as embodying a set of skills that can be 'rented' to employers*". Adding that "*the knowledge and skills that a worker has – which come from academic or other training, including that which comes from experience – generate a certain reserve of productive capital*". Baron and Armstrong (2007) add that "*Human capital is an important element of an organization's intangible assets. Other intangible assets include copyright, consumer relations, brand and image.*"

According to Nerdrum and Erikson (2001), "*intellectual capital is the ability that individuals have to generate added value within an organization.*" According to the bibliographic research carried out, the authors concluded that organizations have a competitive advantage when their assets, capabilities, or processes have specific attributes, making them rare, valuable, and difficult to replicate.

Chatzkel (2004) states that "*the concept of human capital is related to the added value that people add to your organization,*" proceeding "*the reality is that organizations are nothing more than an extension of human thoughts and actions,*" finalizing with "*human capital is the differentiator of organizations, and is the true basis for a competitive advantage.*"

These authors started the movement defending the Theory of Human Capital, which places people at the center of the organization, that is, people are the real source of organizational differentiation, and they are what make organizations efficient, differentiated, unique, and not easily replicable. These factors contribute to the sustainability of organizations.

In addition, with the relevance currently given to "knowledge" in the economy, the management of Human Capital has also attracted the attention of academia and the private sector (Le et al., 2005). Knowledge as an element that enhances the organization's efficiency cannot be ignored, leading organizations to value this knowledge, and to demonstrate an interest in keeping knowledge within their doors, while trying to capture new knowledge abroad. This situation leads to healthy competition between organizations.

This vision of Human Capital as an important asset is not limited to organizations. Kalaitzidakis et al. (2001) studied the relationship between Human Capital accumulation and economic growth, based on Human Capital measures. This question has been raised, analyzed, and promoted by several economists and politicians to defend new proposals, namely the investment in the national education system. The authors concluded that the relationships between human capital and economic growth are very complex, but also that the evidence is consistent in demonstrating the beneficial effects that Human Capital growth at post-primary levels has on per capita economic growth in countries.

A. Wu (2005) divides the organizations' capital in three. Human Capital is characterized as the source of innovation and strategic renewal within an organization, as it incorporates the knowledge, talent, and experience of all employees. Structural capital represents codified knowledge bases, which do not exist in collaborators' minds, e.g., databases, files, organizational routines. Relational capital represents the knowledge existing in the organizational value chain, that is, the knowledge that comes from the relationships that the organization has with suppliers, customers, or any other external entity.

Lin et al. (2017) analyzed 183 participants, from the island of Taiwan and mainland China, and concluded that the management and development of Human Capital are positively related to the value and individuality of employees in both markets, namely the definition of work profiles and training. However, the authors point out contextual (environmental) differences between the two locations, which may influence some relationships. This fact highlights the importance of having individualized human capital management for the organization, aligned with its context, its values, its mission, and its vision.

It is, therefore, relevant for organizations to manage their Human Capital, in order to enhance their efficiency and maintain their competitiveness, by maintaining their Human Capital and attracting new Human Capital.

Human Capital Management—usually abbreviated by HCM or GCM, from the Portuguese Gestão de Capital Humano—is a relatively recent topic, and, as such, there are several definitions and views on the topic. So, it is relevant to look for these various perspectives.

Chatzkel (2004) states that "*human capital management is an integrated effort to manage and develop human capabilities to achieve significantly higher levels of performance.*" Baron and Armstrong (2007) describes Human Capital management as being "*a strategic approach to people management that focuses on issues that are critical to the organization's success.*" Scarbrough and Elias (2002) claim that to date "*the great contribution of human capital management is the definition of a link between human resources and the organization's strategy.*"

Nalbantian et al. (2003) emphasize purposeful measurement as an important feature of HCM. The authors define Human Capital as: "*The warehouse that accumulates knowledge, skills, experience, creativity and other attributes relevant to workers,*" and that the HCM involves "*to put in place the metrics needed to measure the value of those attributes and use those results to effectively manage the organization.*" There is, however, a temptation to limit Human Capital management to measurement, that is, to reduce HCM to numbers. However, it is important to emphasize that the role of HCM is not primarily to measure. HCM's primary role is to create and demonstrate the value that good people and good people management add to the organization. (Baron & Armstrong, 2007).

Donkin (2005) argues that an organization's HCM strengths are visible in three areas: (1) Development and application of relevant, quantitative, and qualitative measures; (2) Gather and interpret results; and (3) Use this information to a strategic advantage. Donkin (2005) adds even that "*organizations that focus this management effort in these areas will be better positioned to align their work policies with their strategic intentions. Good human capital management, therefore, is about learning, understanding, intervening and adjusting.*" In short: measure, analyze, and act—something similar to the PDCA cycle: Plan-Do-Check-Act, characteristic of continuous improvement, and advised in the ISO 9001 Standard.

Therefore, it is important to highlight the planning, measurement, action, and strategic alignment that Human Capital management (HCM) requires. These facts—highlighting the constant search for alignment with the strategy—are very similar to the Balanced Scorecard (BSC). Therefore, IT makes perfect sense to combine the Balanced Scorecard with Human Capital Management, creating the Human Capital Balanced Scorecard.

2.4 Human Capital Scorecard

Le et al. (2005) conducted a literature review and identified three approaches to measuring Human Capital: (1) based on cost; (2) based on salary; and (3) education. These approaches are relevant, but they are focused at the macro level, that is, they are more effective for measuring Human Capital at the level of a country or region, e.g., the Portuguese economy, or the economy of the European Union.

The authors, Becker et al. (2001), argue that HR strategy is the basis of HR's influence. In the book, the model called *High-Performance Work System (HPWS)*. In this model, each HR element is designed to enhance the quality of Human Capital throughout the organization. The quality will depend on the organization's ability to attract and retain talent. HPWS links their selection and promotion decisions to validated competency models and develops techniques that allow for the effective implementation of the strategy in the company and adopts remuneration policies to attract, motivate, and retain the best employees. These HR policies and practices demonstrate a strong alignment of strategy with operational objectives, which brings competitive advantages to the organization.

According to the authors Walker and MacDonald (2001), the HR BSC provides the necessary means to monitor employee indicators, analyze employee statistics, diagnose employee problems, calculate the negative financial impact, prescribe solutions, and track improvements. However, what is really important for top management is to know where these investments are most effective and add value. Should premiums increase or be expanded? Should the benefits be extended? Should training be intensified or more comprehensive? Should recruitment be increased or more thorough? Questions of this type are crucial in an era where capturing and maintaining talent are crucial. At the same time, due to the high competitiveness, all expenses will have an impact on the organization's profits, and, as such, they must be justified to the shareholders or owners of the organization. So, in order to make a sustained decision that can be justified with these stakeholders, the organization should use its HR monitoring strategies.

In order to facilitate the understanding of the Human Capital Balanced Scorecard, Bancaleiro (2007) considers it important to know some aspects that constitute it: the organization's Strategic Plan, the Human Resources Management System, the People Plan, and the Alignment between the People Plan and the organization's strategic plan, which will be presented below.

The organization's Strategic Plan must be characterized by being a simple document, which includes: the main premises on which the plan was based; a description of a scenario that includes the main factors that will influence the future and their impact; the strategic measures to make the best use of this scenario (with the respective objectives to be achieved duly quantified); and the tactics that will enable the strategies, with the respective contingency plans. When defining the business strategy, it is important that the defined objectives are clear.

The Human Resource Management System (HRMS) is the integrated set of Human Resources Management practices, processes, and policies; and it is, within the scope of the Balanced Scorecard, a fundamental component, since it centralizes all the infrastructure necessary to prepare a People Plan (e.g., HR planning, performance management, culture management, talent retention, recruitment, integration, relationships with employees, labor relations, management of working hours).

The People Plan, also known as the Human Capital Strategic Plan, is an important tool, as it defines the priorities and guidelines for the performance of HR professionals; allows alignment between the performance of the HR department and the main objectives of the organization's global strategy (putting the HR area on the same level as other management areas); and sets goals, measuring them. The main advantages of preparing the People Plan are integration (because there is an alignment between the general objectives of the organization, the HR objectives and the objectives of those responsible for the HR department), transparency (which allows all employees to know the objectives of the department), and accountability (ideally each objective should be set to a responsible person, and it is clear to the entire organization who has to deliver that result).

The alignment between the People Plan and the organization's strategic plan—considered by Bancaleiro (2007) as external alignment—seeks to measure the level at which each of the HR objectives / results are aligned with the organization's overall objectives, in order to prevent some results from compromising, or even invalidate, others.

The alignment between the various components of the Human Resources Management System—in which Bancaleiro (2007) calls it internal alignment—translates into determining what the HR area has to deliver in order to achieve global objectives, as well as how the processes that integrate the HRMS are developed; these processes that allow achieving the proposed HR results.

HR Outcome is understood to be what the Human Resources Management System of a given organization, at a given time, delivers to the organization; as well as the indicators related to it (Bancaleiro, 2007). HR Efficiency encompasses a set of tasks (and respective measures) that the SGRH must perform in order to complete the HR Results (Bancaleiro, 2007).

The indicators used for the Human Capital Balanced Scorecard have the following characteristics: strategic, specific, fewer, objective-oriented, dynamic, and future-oriented.

Indicators are strategic, as they select and work with the necessary information, with the objective of measuring the direct or indirect impact on the Balanced Scorecard strategy. These are less numerous, as they are limited to evaluating specific aspects of Human Resources, e.g., employee satisfaction index or the rotation of key people.

3 Research Methodology

In the context of project execution, the applied research methodology was the case study, which allowed the construction and application of the Balanced Scorecard of Human Capital.

To this end, an analysis of the company was carried out in detail, in its entire natural context, based on fieldwork, supported by multiple sources. The purpose of the investigation is always holistic (systematic, broad, and integrated), as it seeks to understand the case. The case study followed the various phases that constitute this methodology. Initially, the objective of the study was defined, which would be the development and application of the Human Capital Balanced Scorecard, recognizing its importance in defining the organization's strategy. Then, the methods for data collection were selected: direct observation, interviews, and document analysis.

In the *direct observation* method, it was possible to be present in the performance of some tasks by some employees of the organization, as well as to know the workplace, its conditions, and the way in which team meetings take place. The *interviews* were carried out as a way of getting to know the organization's strategy, what is expected from Human Capital, and the benefits given to employees who are aligned with the strategy. In the document analysis method, documents were analyzed, namely the communication of annual results and the communication of the strategy to the various levels of the organization, in order to better understand the relationship and form of communication between top management and its employees.

After the first analysis, the current performance indicators were identified and understood, in order to understand the connection between them all; this connection must be coherent and allow for better assessment and monitoring of the organization's current direction.

A qualitative methodology was adopted. Initially, a mixed methodology, of a qualitative and quantitative nature, was planned; however, due to the SARS-CoV-2 pandemic, the quantitative component—designed through employee surveys—was abandoned.

As mentioned above, the SARS-CoV-2 pandemic forced a rethinking of the approach, namely it was not possible to include a quantitative methodology. This change forced new forms of collaborative work, communication, and sharing. Such changes forced top management to rethink and restructure various aspects of the organization. These changes created an additional workload on the organization's employees, not allowing the administration of a questionnaire to them.

3.1 Profiling the Studied Organization

The consultant under study is made up of 22 employees, approximately two thirds of whom are female. All employees are licensed, most have post graduation or master's degrees. It is a young team aged between 20 and 44 years. The organization under study is not departmentalized, in contrast to many private and public organizations. This means that there is no typical HR department, in charge of all human resource management (HRM) tasks. The organization has a matrix structure, divided into three areas: financial, strategic, and Human Capital. These three areas coincide with the three areas of intervention of the organization in terms of consultancy.

Of these three areas of intervention, it is Human Capital that affects the group's human resource management. This area is responsible for the management and operation of all HR processes, across all areas. This approach to management is common to all areas, applying standard processes, to ensure homogeneity throughout the organization.

3.1.1 HR Role

The Human Capital area, as mentioned above, performs all the functions related to HRM for the various areas of the organization. These functions are: structure management of all work-family reconciliation practices; absence management; diagnosis of the training plan; training plan management; training impact assessment; development plan; definition of the goals, integration and reception plan; organizational/functional climate assessment; management of functions; category analysis; career advancement; wage policies; and code of conduct (through the regulation).

The task that is usually assigned to the HR Department, but in the organization under study, is not carried out by the Human Capital area is the processing of salaries. This task is carried out by the finance department.

The top management of the organization under study considers that all human resources practices must be considered as a whole, as they are all necessary and strategic, that is, they are interconnected and have consequences in various aspects of the organization. All HR practices implemented resulted from the pursuit of the organization's strategy, and all these practices are articulated with each other. The organization considers it crucial that the entire team identify with its mission, vision, and values, as this is the only way that Human Capital will be aligned with the organization's strategy.

3.1.2 Communication

Communication within the organization is carried out in different ways, in order to promote better fluidity. A global meeting takes place every year, with all areas of the organization, where the results of the exercise are shared, as well as the strategy for the following year. Monthly objectives meetings are also held, where monthly objectives by area are defined, communicated, and evaluated. Weekly, the coordinators of each area meet with the management. In addition to these meetings, communication is also carried out in a more informal way: via email or directly, where there is the possibility of clarifying doubts or questions about any matter of the organization, since the administration is often present at the headquarters. Additionally, the organization designed two computer work tools, which aim to share information more quickly and efficiently, not only within the team, but also with the customer; also allowing to structure and present the work processes in each project, which also promotes better time management.

3.1.3 Recruitment

The Human Capital area performs recruitment not only for the organization under study (in its various areas), but also for its clients, since the organization provides HRM services, namely recruitment and selection. The organization under study has several partnerships with Higher Education Institutions. These recruitment sources were chosen because they meet the profile defined by the organization for the various functions, but also because of their geographic proximity to the organization's facilities.

After recruitment, the first stage of employee integration consists of following a plan that is already defined by the organization called "Reception." It is assigned a mentor to the new employee, who will follow up throughout the organization's first year. Also, during this phase, workshops are held in the areas of communication and relationship, shake your team (an activity to encourage teamwork), and the balance of time; in order to familiarize new employees with organizational values.

3.1.4 Performance Evaluation

The mentor assigned to the new employee will be responsible for conducting a quarterly follow-up with the mentee, to understand their evolution, the difficulties experienced, and define the goal plan for the following quarter, considering the level of knowledge current and the development to be carried out in functional terms.

3.1.5 Career Management and Promotions

About career management and promotions, the organization is based on a plan of objectives that is based on the competence profile defined for each function, and on the development of these competences by levels. This plan of objectives is presented annually and evaluated every six months. This always covers hard-skills and soft-skills. More practical objectives are also defined, e.g., in terms of managing work times. The biannual evaluation meeting is where the employee has the opportunity to carry out their self-assessment; to know what the evaluation of your leadership is, considering the results achieved; there is a sharing of information that allows for some points to be discussed; or possibly understand the reasons why certain goals are not being achieved. At the end of each year, a general meeting is held, where the assessment of the year is communicated, and a reassessment of the aforementioned plan of goals.

3.1.6 Monetary and Non-Monetary Bonuses Policy

The bonus policy can be monetary or non-monetary, according to performance or specific need. At the annual strategy meeting, it is communicated to all employees whether the defined objectives have been achieved. If so, there is a distribution of the result, which is reverted into an annual monetary bonus. The organization offers other benefits to its employees, such as: extra day of vacation on their birthday; co-payment of health insurance; individual development plan (the employee himself can propose an action that is of interest to him and provided that it is in line with the profile defined for his role); more than 40 h of annual training; partner discount card (usually customers); team building actions; co-payment of tuition fees in master's courses; internal qualifying sessions; bet on ergonomics at work; and gift card in maternity.

4 Analysis and Discussion

This chapter portrays the implementation phase in the organization of the Human Capital Balanced Scorecard. The first subchapter depicts the initial diagnostic phase, presenting the analysis of the current state of the organization, its opportunities for improvement and ambitions. In the second subchapter, the SWOT analysis carried out on the organization is presented. In the following subchapters, the work carried out to reach the Human Capital Balanced Scorecard is presented. In these subchapters, the strategic objectives are listed, summarized in the People Plan, and later detailed through the specific goals, which form the Human Capital Balanced Scorecard.

4.1 Diagnosis

The organization's diagnosis was carried out through meetings with one of the organization's co-CEO. These meetings focused mainly on understanding the company's strategy, as well as understanding the interactions between the various areas of the organization, and even the challenges that the SARS-CoV-2 pandemic brought to the organization, people, and business.

4.2 SWOT Analysis

In order to better characterize the organization, to reflect on its current situation, and to structure this information, a SWOT analysis was performed. However, the main objective of carrying out the SWOT analysis is to analyze whether the strategy currently defined by the organization is in line with the current situation, from an internal point of view (Strengths and Weaknesses), and from an external point of view (Opportunities and Threats), or if the strategy needs adjustments. With regard to the internal environment, the following *strengths* were identified: concern with Human Capital, insofar as there is a concern on the part of the company in reconciling work and family, distinguishes the company from others in the sector, encouraging retention of talent. The organization has a young and dynamic team, with new perspectives and always attentive to change. The company systematically invests in innovation, through the implementation of projects that aim to improve some aspect of a client's internal work or processes, fostered through transparent incentive policies. The organization offers a range of services, which aim to respond to the various needs of Portuguese SMEs (Small and Medium Enterprises), which may represent the obvious choice for a client, after having contracted their first service with them—in addition to make the team multidisciplinary. Its geographical locations in cities with a vast business fabric offer an advantage in order to reach a greater number of customers.

Still, from an internal perspective, the organization has two weaknesses. The first refers to the absence of an IT (Information Technology) consulting team, which enables a diagnosis, evaluation and offer of IT solutions to customers, with a view to improving processes, management, and information flow. This weakness is crucial in an era where IT makes companies value, makes processes more efficient, and facilitates the connection with the end customer. The second weakness points to the absence of an internal tool for measuring the organization's strategic alignment with the operational activity, which allows management to assess whether the organization's daily operations are in fact contributing to the strategy defined by it. This piece of work responds to this weakness, in order to provide the organization with a Human Capital Balanced Scorecard. As for the external environment, they were identified as opportunities: in the post-pandemic, in order to improve its sustainability, customers should aim to improve their efficiency, which could result in an increase in projects for the company under analysis. Through their previous experience, the emergence of European post-pandemic financing, with the aim of stimulating the economy—could lead to an increase in projects and clients. By contrast, the following threats were considered: the uncertainty of the duration of the pandemic, which will make it difficult to acquire new customers, the implementation of current customer projects, and the investment in new projects by customers. The economic contraction, resulting from the pandemic, could reduce investments by customers.

4.3 Human Resource Management System

The human resource management system summarizes the inputs (organizational needs and human aspirations) and the outputs (organizational effectiveness and individual growth).

Bancaleiro (2007) states that the human resources management system comprises all the infrastructure necessary to carry out the People Plan. Reinforcing that this

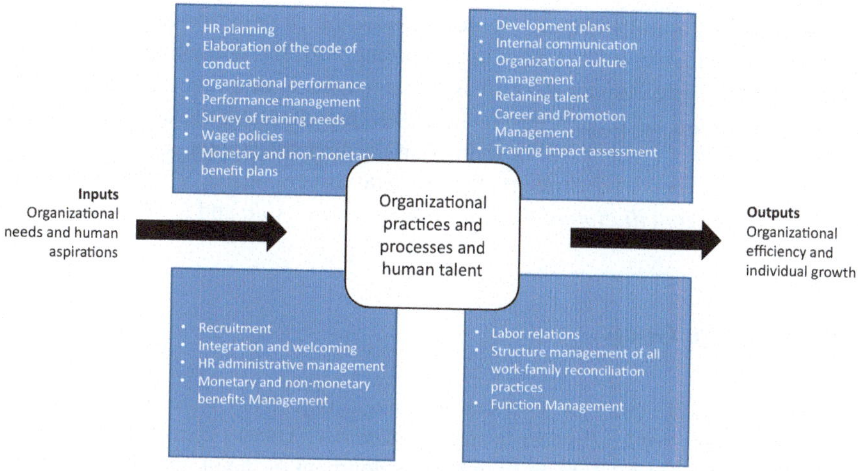

Fig. 1 Human resource management system, adapted of Bancaleiro (2007)

model conveys a dynamic idea, which should be the basis of any management system, this model presents the inputs and outputs of the human resources management system, conveying the idea that the set of instruments that make up the system has an operational objective; this distinguishes between processes with more operational characteristics and more strategic ones; the model conveys an idea of integrating processes that complement and enhance each other—which is a fundamental characteristic of any system.

The model allocates the main processes of the human resource management system across the four sectors (divided into quadrants), based on criteria of greater or lesser strategic weight, combined with greater or lesser connection to people.

The upper left quadrant demonstrates the fundamental processes for greater connection to the business. In the upper right quadrant are the processes most linked to organizational and personal change. By contrast, in the lower right quadrant are the more operational processes that are linked to organizational relationships, while in the lower left quadrant are those processes that have a more every day and more technical component. Figure 1 demonstrates the human resource management system designed for the organization under analysis. This model was created based on the analysis that was made to the organization and is an adaptation of Bancaleiro (2007).

4.4 *Strategic Drivers*

In developing a Balanced Scorecard, it is crucial to understand the organization's strategy. This is because it is the organization's strategy that feeds the Balanced Scorecard, as shown in Fig. 1. The organization's strategy must be defined in precise terms, and understandable by all its employees.

The Human Capital Balanced Scorecard uses the strategic drivers as the organization's governance, with a view to complying with the defined strategy. These drivers should guide Human Resources Management in defining its objectives, to work together with the other areas, in an aligned manner, toward the organization's strategic objectives - which will be defined in the next subchapter.

In line with point 3.1.3 Strategy, the strategic drivers of the organization under study are: (1) *Reconciliation of the work-family relationship* and (2) *Expansion.*

4.5 Strategic Goals

In line with the organization's strategy, strategic objectives were defined in four areas: *Strategic, People, Processes* and *Operational.* These objectives represent a macro perspective and will be used in the development of the organization's People Plan, and in the definition of specific objectives; the last ones being from a micro perspective.

Figure 2 outlines these strategic objectives. At the center is the mission of the consultancy's Human Capital area: to achieve excellence in attracting, developing, and motivating people, and in managing the organization. From the center come the

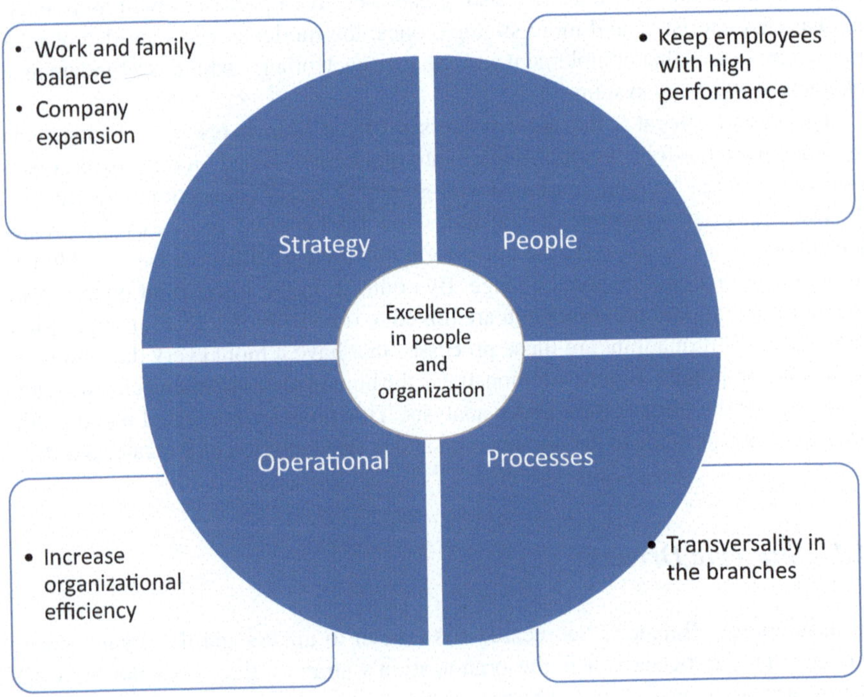

Fig. 2 Strategic goals of organization

four areas: *Strategic, People, Processes* and *Operational*. Each area displays, at the four ends of the figure, its strategic objectives.

The *Strategic* area has two strategic goals, as these are considered by top management as essential for the organization's long-term sustainability. The remaining areas: *People, Processes* and *Operational*, have a strategic goal each.

The *Strategic* area includes expansion and the work-family balance. The first is relevant to improve the organization's profitability, since the organization's processes and know-how can be scalable to other customers, with expansion being the most natural way to make the investment already made profitable and to reach new customers. The second strategic goal, the work-family balance, is crucial to maintain the motivation of human resources in carrying out their tasks and promote focus during the development of work tasks; but also, in order not to force employees to choose between a professional career versus family life—usually placed on the shoulders of female employees—since harmony between both is the organization's goal.

In the *People* area, the maintenance of high-performance employees is the strategic objective, in order to retain the knowledge built up over the years and continue to have good efficiency indices in the development of work tasks. This balance between effectiveness and efficiency will keep the organization competitive. Through this retention of high-performing employees, the organization will (a) save on training new employees, who might be needed to replace those who may eventually leave; and, at the same time, (b) through the Maintenance of high-performing employees, it may be possible to train new employees with mentors who motivate, inspire, and promote good work practices. These new employees may be important to the expansion strategic objective, presented above.

The *Processes* area includes the strategic objective of transversality of processes in the various branches. This homogeneity of processes between the organization's branches is intended to promote connection and communication between employees in the same area but linked to different branches. That is, this objective aims to promote interaction between people who perform the same task in different branches, promoting the discussion of ideas and the sharing of experiences, to increase global knowledge, reduce potential errors, and strengthen processes.

In the *Operational* area, the strategic objective is to strengthen organizational efficiency, to improve the effectiveness of processes in the various branches (since they are known, tried, and improved in all branches), it simplifies the movement of employees between branches (which favors efficiency in the development of work), and facilitates collaborative work between employees of the various branches (e.g., if they work on a larger project). All these points are relevant to the organization's efficiency, streamline the management of human resources, and simplify the management of holding.

4.6 People Plan

In line with the strategic objectives presented, the organization's People Plan was developed, adapted from Bancaleiro (2007). The People Plan has at its core—like the strategic objectives in Fig. 2—the mission of the consultant's Human Capital area: to achieve excellence in attracting, developing and motivating people, and in the management of the organization. From this maximum objective come the specific objectives of the four areas. In the outer circle of the People Plan, there are specific goals, which derive from, and help to achieve, each specific goal.

Bancaleiro (2007) represents the People Plan in the circular shape visible in Fig. 3. Such representation is positive from the point of view of the organization of areas and objectives, as well as being visually more appealing. Each strategic goal gives rise to a critical result in the People Plan's human resources management developed for the company. That is, each strategic objective must achieve a result, which is critical in HRM, to comply with the strategy defined by the organization for its HR.

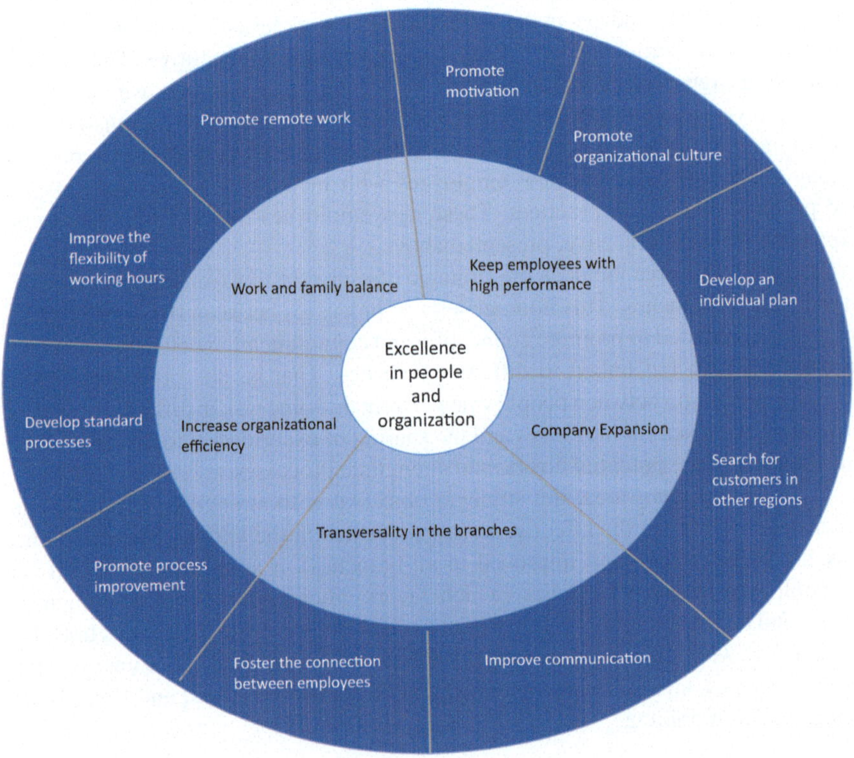

Fig. 3 People plan proposal

4.7 People Plan Critical Human Resources Results

This topic presents (a) the critical human resources results of the company's People Plan, (b) the relationship between the strategic objective and its specific objectives, and (c) the performance indicators to be used to assess the status of each specific objective, and consequently the critical result.

Each critical result comes from a strategic objective, which is subdivided into specific objectives responsible for measuring efficiency in human resource management, through the assigned indicators.

The *Strategic* area was divided into two strategic objectives, as they are two crucial components for the sustained future of the organization; then give rise to two critical results. The first (1) Work-family balance, is something the organization already aimed for, but it has increased its relevance during periods of confinement caused by the SARS-CoV-2 pandemic.

HR Outcome 1—Promote work-life balance Medidas de Eficiência RH:

- Improve schedule flexibility;
- Encourage teleworking.

Indicators:

- Average absences per month;
- Type of absences;
- Work-family balance index;
- Time exemption index;
- Teleworking index.

To respond to this critical result, the following specific objectives were defined: (a) *Improve flexible hours* and (b) *Encourage remote work*. The specific objective (a) points to the answer to the growing need to adapt working hours to the fluctuating needs of personal life; that is, allowing employees to adapt their working hours to their specific needs, by increasing the flexibility of each employee's working hours. The need for flexibility may come from a call to the child's school, or a doctor's appointment. This flexibility can also be reflected in the improvement of the employee's quality of life, as it can be used to offset the employee's travel to and from the company, with the busiest times in their city—reducing daily travel stress, and reducing the time lost by the employee in traffic, being able to use this time either in the company or in their personal life.

Through the specific objective (b) *Encourage remote work*—also known as telework—the employee will not travel to and from the company's premises on several days of the week, being able to carry out their work from their home. This dispensation may be interrupted occasionally due to a specific need to visit a client's premises, or to be present at a team meeting at the company. To measure the fulfillment of this objective, the indicators will be used: *Average absences per month, Type of absences,* and *Work-family balance index*. The first two point to a direct advantage of this work-family balance for the organization: a reduction in the number of absences, i.e., if there is flexibility on the part of the organization toward the personal lives of its employees, absences due to monitoring the lives of descendants or for reasons of non-urgent consultation may reduce. The *Work-Family Reconciliation Index* indicator measures the perception of employees in relation to reconciling their professional work with their personal life, one of the objectives and advantages of teleworking.

Remote work is partly linked to the argument of the previous specific objective, and can also reduce daily travel stress, and reduce the time lost by the employee in traffic, being possible to use this time either at work or with their family. In addition, this excuse for travel will have a positive effect on reducing the expenses associated with the employee's daily travel to and from the company's premises. The indicators chosen for this objective were: *Time exemption index and teleworking index*, both suggested in the scope of this work. The first indicator, exemption from working hours index, measures the percentage of employees with an exemption from working hours in relation to the total number of employees in the organization. This time exemption allows you to follow a schedule that is more conducive to your personal life, e.g., to be able to take your children to school, or to reconcile your schedule with that of your spouse. The second indicator demonstrates the percentage of employees in the organization who are working remotely, in relation to the total number of employees in the organization.

HR Outcome 2—Company Expansion

HR Efficiency Measures:

- Search for customers in adjacent regions.

Indicators:

- Number of new customers.

The second strategic objective (2) *Expansion*—still placed in the Strategic area, establishes as a specific objective: (c) *Search for customers in adjacent regions.* This objective intends to take advantage of the population density, and consequent business density, of the Minho region, and look for clients in regions close to the consultant's current facilities, which may eventually result in the opening of branches in other cities in the region. The chosen indicator was the *Number of new customers*, in order to calculate the number of new clients who signed consulting contracts with the organization. This expansion aims to make the investment made in the organization's human resources profitable and ensure the sustainability of the business.

HR Outcome 3—Keeping high-performance employees

HR Efficiency Measures:

- Develop an individual career management plan.
- Promote organizational culture.
- Promote motivation.

Indicators:

- Career and compensation management index.

- Training and skills development index.
- Number of activities that promote organizational culture.
- Organizational engagement and commitment index.
- Organizational happiness index.

In the *People* area, the organization's strategic objective is (3) *Keeping employees performing at a high level.* To respond to it, the following specific objectives were established: (d) *Develop an individual career management plan*; (e) *Promote the organizational culture; and* (f) *Promote motivation.*

The first specific objective (d) *Develop an individual career management plan*, points to the individualization of career management, that promotes individualized management that is closer to the employee, as well as commits the organization to an analysis of the career of each employee. This point is intended to improve the individual knowledge of each employee, act proactively and accordingly on a change in employee performance, promote their motivation, and analyze possible opportunities for improvement. As indicators, the following were chosen: *Career Management and Remuneration Index,* and *Training and Skills Development Index*, both owners of the organization, that is, they are indicators that result from the combination of other indicators, whose formula is the organization's intellectual property. The first is linked to the individual management of the career development of each worker, as well as the remuneration associated with that evolution; while the second is linked to the training given to each employee, in line with the development of skills necessary for their individual career evolution within the organization.

The second specific objective, (e) *Promote organizational culture*, has two bases: (1) promote the organization's culture among all employees, promoting the focus on the consultant's strategic objectives, maintaining its values; and (2) promote the connection between employees, who will all be aligned with the organizational strategy. This point is relevant for the company's expansion, because, with the fragmentation of the organizational structure, it is important to promote the organization's culture with all its human resources. The chosen indicator was the *organizational happiness index*, organization owner, who provides the organization with feedback on the happiness of employees within the organization.

The third specific objective (f) *Promote motivation*, is intended to encourage the motivation of employees, so that it remains at levels considered adequate for the performance of the organization. That is, that the motivation to work for the organization is at the levels necessary for the organization to be able to achieve the established goals. In addition, promoting employee motivation helps maintain employees in the organization. This promotion also has the advantage of attracting new talent to the organization. The chosen indicators were the *Number of activities that promote organizational culture,* and *Organizational involvement and commitment index*. The first indicator was suggested in the development of this work, and it counts the number of activities carried out by the organization to promote the organization's culture, e.g., the annual result sharing event. The second indicator is the organization's owner and intends to measure the level of involvement of the organization's employees in the development of the organization's activities.

HR Outcome 4—Crosscutting processes in the branches

HR Efficiency Measures:

- Facilitate communication between employees with the same activity.
- Foster the connection between employees who carry out the same activity in different branches.

Indicators:

- Communication index.
- Cooperation and teamwork index.

In the *Processes* area, the strategic objective is (4) *Crosscutting in branches.* This points to the use of common processes in the various branches, with the aim of improving their effectiveness and promoting the efficiency of the same activity in the different branches. At the same time, it is intended that the know-how is transmitted between the various branches, that is, the knowledge acquired in one branch should be shared in other branches, in order to collectively improve the organization's processes.

To achieve this strategic objective, two specific objectives were defined: (g) *Facilitate communication between employees with the same activity*, and (h) *Foster the connection between employees who carry out the same activity in different branches.*

The specific objective, (g) *Facilitate communication between employees with the same activity*, points to facilitating the exchange of information between employees who carry out the same activity, whether they are in the same building or in different branches; in order to improve the flow of information, and thus improve the effectiveness and efficiency of the organization's processes. The indicator for this specific objective is the organization's proprietary *Communication Index*, which provides feedback on the opinion of internal communication collaborators.

The specific objective, (h) *Foster the connection between employees who carry out the same activity in different branches*, aims for a better interpersonal relationship between employees who perform the same function in the various branches of the organization. With this connection, it is intended to encourage the sharing of knowledge and tools among employees who perform the same function in the various branches, improving the organization's effectiveness and efficiency. The chosen indicator was the *Cooperation and teamwork index*, owner of the organization, which measures how easy it is for employees to cooperate and collaborate with other members of the organization.

HR Outcome 5—Reinforce the efficiency of organizational processes

HR Efficiency Measures:

- Promote ideas for continual improvement.
- Develop standard processes.

Indicators:

- Rate of completion of suggested ideas.
- Implementation rate of ideas for changing service and/or organizational processes.

The strategic objective is (5) *Strengthen organizational efficiency*. Its main goal is to improve the efficiency of organizational processes. To achieve this strategic objective, two specific objectives were defined: (i) *Promote ideas for continuous improvement* and (j) *Develop standard processes.*

The specific objective (i) *Promote ideas for continual improvement*, aims to internally promote the suggestion of ideas to improve the organization's processes. This will have as agents of change the organization's own employees, who carry out the organizational processes daily, and who have first-hand access to the outputs of these same processes, the effort necessary to obtain them, and the customer's reaction to them. So, all employees are important pieces in the continuous improvement of organizational processes. The indicator defined to measure this specific objective was the *Rate of completion of suggested ideas*, which measures the percentage of implementation of ideas suggested by employees.

The objective (j) *Develop standard processes* points to the definition of generic processes, which are adaptable to the different realities of the present and future branches, which may have different dimensions and values. Through standard processes, the improvement of a process will have positive repercussions in the other branches, being able to improve the effectiveness and efficiency of the organization transversally to its branches. The indicator established for this specific objective was the *Implementation rate of ideas for changing service and/or organizational processes*, which measures the percentage of implementation of ideas suggested by employees with a view to improving the services provided by the organization, and the processes used by the organization.

4.8 Human Capital Balanced Scorecard

With the strategic drivers defined, and the critical results of the People Plan presented, it is now possible to proceed to the preparation of the Human Capital Balanced Scorecard, where the strategic drivers are associated with the performance indicators, to verify the alignment between the results of HR measured by performance indicators, and the strategic drivers defined by the organization.

The Table 1 presents the proposed Human Capital Balanced Scorecard developed for the company. The leftmost column contains the strategic drivers. These are linked to critical results, closely linked to the five strategic objectives, which are subdivided into specific objectives. Each specific objective has one or more performance indicators associated with it. Each chosen performance indicator presents its goal, i.e., the measurable objective to be achieved in each indicator, to measure the organization's governance.

Table 1 Proposed Human Capital scorecard

Strategic drivers	Critical results	Assessment indicator	Goal	Frequency	Responsible
Work and family balance	Promoting a balance between professional and personal life	Average absences per month	5%	Monthly	HC Responsible
		Type of absences	Occasional absences	Monthly	HC Responsible
		Time Exemption Index	1% of employees	Semester	Team Manager
		work-family balance index	3,2/4,0	Semester	Team Manager
		Telecommuting index	2% of employees	Semester	Team Manager
Company Expansion	Company Expansion	number of new customers	20/year	Yearly	CEO
	Keeping high-performance employees	Career and Compensation Management Index	3.3/4.0	Yearly	Area responsible member
		Training and skills development index	3.3/4.0	Yearly	Team Manager
		Number of activities that promote organizational culture	2 years	Yearly	CEO
		Organizational involvement and commitment index	3.4/4.0	Yearly	Team Manager
		Organizational happiness index	3.0/4.0	Yearly	Team Manager
	Crosscutting processes at branches	Communication index	3.3/4.0	Semester	Hierarchical superior
		Cooperation and teamwork index	3.3/4.0	Semester	Branch director

(continued)

4.9 Communication and Implementation

Communication and implementation are a crucial phase, as it will allow the organization's management and other areas to be aware of the strategic commitments that have been made in terms of HR, and how these will be measured (Bancaleiro, 2007).

Table 1 (continued)

Strategic drivers	Critical results	Assessment indicator	Goal	Frequency	Responsible
	Strengthen the efficiency of organizational processes	Rate of completion of suggested ideas	90%	Quarterly	CEO
		Realization rate of ideas to change service and/or organizational processes	95%	Quarterly	CEO

It is important to invest in internal communication, as the success of its application and maintenance depends on the collaboration of the entire team. Furthermore, it is important that it is correct, reliable, and communicated with an adequate explanation and a guide for implementation and action (Baron, 2011). Communication should be done from the top down, since it is about decisions on strategic issues, which are fundamental for the organization.

4.10 Communication Periodic Evaluation

After implementation, it is equally important to periodically evaluate and rectify the BSC-CH (Bancaleiro, 2007; Baron, 2011). This is because it is through the Human Capital Balanced Scorecard that the Human Capital manager obtains indications that allow him to take timely measures to reinforce or correct the objectives defined in the People Plan.

5 Final Remarks

The Balanced Scorecard is increasingly recognized as a crucial tool to assess the alignment of the organization's daily tasks with the organizational strategy, quantitatively measuring the possible mismatch between the two parties, allowing to act proactively in the event of any deviation. However, this exercise requires in-depth knowledge of the organization, both in terms of its strategy (including the mission, vision, and values that feed it), but also in terms of its organizational and business processes.

To get to know the organization, a diagnosis was carried out. This resulted in the SWOT analysis of the organization and the presentation of the Human Resources Management System. Together with the company's administration, the strategic drivers responsible for the governance of the organization were defined. With these, it was possible to develop the strategic objectives, which although macro, share the focus of the strategic drivers. From these strategic objectives, it was possible to build the organization's People Plan, which adds the specific objectives. Then, the critical HR results were presented, which are accompanied by the respective indicators.

In the development of this investigation, the creation of three performance indicators was suggested: *teleworking index*, which represents the percentage of employees working remotely in relation to the total number of employees in the organization; the *Time Exemption Index*, which shows the percentage of employees with exemption from working hours in relation to the total number of employees in the organization; and the *Number of activities that promote organizational culture*, which counts the number of activities developed to promote the organizational culture. Such initiatives allowed the construction of the Human Capital Balanced Scorecard of the consultant under analysis, based on the defined strategy, and serving the critical results of HR.

When performing the external alignment framework, it was verified that the two strategic drivers *Conciliation of the work-family relationship* and Expansion may have a negative conflict with critical results *Expansion of the company and promoting a balance between professional and personal life*, respectively. This is due to the fact that the investment in the company's expansion requires an additional commitment from the organization's employees, which may jeopardize the promotion of a better relationship between the professional and personal life of the employees involved.

In addition, the two specific objectives of the critical result *Promoting a balance between work and personal life*—(a) *Improve working hours flexibility and* (b) *Encourage teleworking*—may complicate the strategic driver *Expansion*, since, respectively, it desynchronizes the working hours of employees and promotes work in a different place from the new branch. Misalignment by itself creates some difficulties, as a critical result may be positively influencing one strategic driver, while negatively influencing another strategic driver. In this case, there is a direct conflict between two critical results and two strategic drivers in the external alignment. That is, to positively influence one of the strategic drivers through a critical result, the other strategic driver will be negatively influenced. The company must prioritize one of the strategic drivers over the other. This is because the two strategic drivers are in direct conflict.

Through the internal alignment framework, it appears that a way to achieve the critical result *Strengthen the efficiency of organizational processes* is to invest in the HR Efficiency Process *Remuneration and benefits*. That is, through this HRM process, employees who suggest an improvement in organizational processes with a positive impact on their efficiency may be rewarded with a direct benefit. Also in the organization's internal alignment, it was verified that the HR *Communication Systems* efficiency process has a positive impact on all critical results; because the communication of the strategy is essential in the organization's alignment, so that each employee can know their role in this global strategy.

Regarding the limitations of applied research, confidentiality issues and the influence of the SARS-CoV-2 pandemic made it impossible to have a more direct contact with the company under study. Due to the limitations presented and the shorter period of the investigation process, it was not possible to verify the evolution of the various performance indicators defined, to identify opportunities for improvement.

References

Bancaleiro, J. (2007). *Scorecard de Capital Humano*. Editora RH.

Baron, A. (2011). Measuring human capital. *Strategic HR Review, 10*(2), 30–35.

Baron, A., & Armstrong, M. (2007). *Human capital management: achieving added value through people*. Kogan Page Publishers.

Becker, B. E., Huselid, M. A., Huselid, M. A., & Ulrich, D. (2001). *The HR scorecard: Linking people, strategy, and performance*. Harvard Business Press.

Bieker, T. (2002). Managing corporate sustainability with the balanced scorecard: Developing a balanced scorecard for integrity management. oikos PhD summer academy 2002—"Sustainability, Corporations and Institutional Arrangements", 2–20.

Chatzkel, J. (2004). Human capital: The rules of engagement are changing. *Lifelong Learning in Europe, 9*(3), 138–145.

Donkin, R. (2005). *Human capital management: A management report*. Croner.

Ehrenberg, R. G., & Smith, R. S. (1997). *Modern labor economics: Theory and public policy* (6st ed.). Taylor & Francis.

Kalaitzidakis, P., Mamuneas, T. P., Savvides, A., & Stengos, T. (2001). Measures of human capital and nonlinearities in economic growth. *Journal of Economic Growth, 6*, 229–254.

Kaplan, R. S., & Norton, D. P. (1992). The balanced scorecard: Measures that drive performance. *Harvard Business Review, 83*(7), 172.

Kaplan, R. S., & Norton, D. P. (2000). Putting the Balanced Scorecard to work. *Focusing Your Organization on Strategy—with the Balanced Scorecard, 2*, 2–18.

Le, T., Gibson, J., & Oxley, L. (2005). Measures of human capital: A review of the literature. New Zealand Treasury Working Paper No. 05/10.

Lin, C., Yu-Ping Wang, C., Wang, C.-Y., & Jaw, B.-S. (2017). The role of human capital management in organizational competitiveness. *Social Behavior and Personality: an International Journal, 45*(1), 81–92.

Marshall, A. (1982). *Princípios de Economia*. Abril Cultural.

Mendes, P., Santos, A. C., Perna, F., & Ribau Teixeira, M. (2012). The balanced scorecard as an integrated model applied to the Portuguese public service: A case study in the waste sector. *Journal of Cleaner Production, 24*, 20–29.

Nalbantian, H., Guzzo, R., Kieffer, D., & Doherty, J. (2003). *Play to Your Strengths: Managing Your Internal Labor Markets for Lasting Competitive Advantage* (1st ed.). McGraw-Hill Education.

Nerdrum, L., & Erikson, T. (2001). Intellectual capital: A human capital perspective. *Journal of Intellectual Capital, 2*(2), 127–135.

Pinto, F. (2007). *Balanced Scorecard Alinhar Mudança, Estratégia e Performance nos Serviços Públicos de Francisco Pinto* (1st ed.). Edições Sílabo.

Scarbrough, H., & Elias, J. (2002). *Evaluating human capital*. CIPD Publishing.

Silva, N. J. M. P. de S. e. (2010). *Gestão estratégica do crescimento económico em Portugal*. Vida Económica.

Smith, R. F. (2007). Business process management and the balanced scorecard: using processes as strategic drivers. In Balanced scorecard step-by-step (pp. 127–165). John Wiley & Sons, Inc.

Walker, G., & MacDonald, J. R. (2001). Designing and implementing an HR scorecard. *Human Resource Management, 40*(4), 365–377.

Wu, A. (2005). The integration between balanced scorecard and intellectual capital. *Journal of Intellectual Capital, 6*(2), 267–284.

Wu, H. (2012). Constructing a strategy map for banking institutions with key performance indicators of the balanced scorecard. *Evaluation and Program Planning, 35*(3), 303–320.

Index

C. Machado (ed.), *Challenges and Trends in Organizational Management and Industry*, Management and Industrial Engineering,
https://doi.org/10.1007/978-3-030-98048-1

The manufacturer's authorised representative in the EU is Springer Nature Customer Service Centre GmbH, Europaplatz 3, 69115 Heidelberg, Germany. If you have any concerns regarding our products, please contact ProductSafety@springernature.com

Printed and bound by CPI Group (UK) Ltd, Croydon, CR0 4YY
22/09/2026
02220961-0013